I0043718

J. G. S. Drysdale

Protoplasmic Theory of Life

J. G. S. Drysdale

Protoplasmic Theory of Life

ISBN/EAN: 9783337095512

Printed in Europe, USA, Canada, Australia, Japan

Cover: Foto ©berggeist007 / pixelio.de

More available books at **www.hansebooks.com**

THE

PROTOPLASMIC THEORY OF LIFE.

BY

JOHN DRYSDALE, M.D. Edin., F.R.M.S.

ONE OF THE EDITORS OF FLETCHER'S "PATHOLOGY;"
AUTHOR OF "PHYSIOLOGICAL ACTION OF KALI BICHROMICUM,"
AND OF "LIFE AND THE EQUIVALENCE OF FORCE;"
CO-AUTHOR OF "HEALTH AND COMFORT IN HOUSE-BUILDING."

LONDON:

BAILLIÈRE, TINDALL, AND COX.

20, KING WILLIAM STREET, CHARING CROSS.

PARIS: BAILLIÈRE. MADRID: C. BAILLY-BAILLIÈRE.

1874.

PREFACE.

THE subject of this work was the theme of an inaugural address to the Microscopical Society of Liverpool, delivered by me as President for the year 1874. My attention had been directed to the subject for several years, and materials accumulated for the treatment of it far beyond what could be given in the compass of a lecture. These materials have been, to some extent, used in the Second Part, and the rest were intended for the Third Part of " Life, and the Equivalence of Force," a work in which I have endeavoured to bring before the notice of biologists the remarkable anticipation of certain recent views on the nature of life and other physiological questions, contained in the, partly posthumous, works of Dr. John Fletcher, of which I am the sole surviving editor. It has seemed, however, better to publish what refers to this subject complete in itself, leaving my hands free for other matters, especially the stimuli: hence the present book. As no claim to original discoveries is here put forward,

I have preferred giving extracts or full analyses of the writings of the original observers to whom the building up of the Protoplasm theory is due, rather than writing a compendium which would most probably fail to give so accurate or interesting an account of them. Also, as this work is addressed to men of general culture in science rather than to those technically educated, I have entered on the general physiology of some parts of the subject more fully than would be required by the latter. But by these means it is hoped that men of general scientific culture may have the opportunity at hand of judging of the important theory attempted, however imperfectly, to be set forth in this book, viz., that every action properly called vital, throughout the vegetable and animal kingdoms, results solely from the changes occurring in a structureless, semifluid, nitrogenous matter now called Protoplasm.

LIVERPOOL, *October*, 1874.

TABLE OF CONTENTS.

CHAPTER V.

CHAPTER VI.

CHAPTER VII.

CHAPTER VIII.

CHAPTER IX.

CHAPTER X.

CHAPTER XI.

CHAPTER XII.

THE PROTOPLASMIC THEORY OF LIFE.

CHAPTER I.

INTRODUCTION, AND FLETCHER'S THEORY OF ONE ONLY LIVING MATTER.

WERE it possible even to do so in less than a volume, it would be tedious and unprofitable to go over the history of all the theories of life, so I will begin at the point where it became clearly apparent that all the varieties of opinion might be summed up under two heads—

1. Those which require the addition to ordinary matter of an immaterial or spiritual essence, substance, or power, general or local, whose presence is the efficient cause of life ; and

2. Those which attribute the phenomena of life solely to the mode of combination of the ordinary material elements of which the organism is composed without the addition of any such immaterial essence, power, or force.

Up to the year 1835, the balance had been inclining against the hypothesis of a vital principle, at least in the crude form hitherto predominant, but the minds of

1

physiologists were far from clear, and the ideas of some central vital influence, which ruled over all local actions, or even furnished vital influence to them, were still in the ascendant.

In 1835, Joh. Müller commenced an essay on "Organism and Life" with the following words of Kant: "The cause of the particular mode of existence of each part of a living body resides in the whole, while in dead masses each part contains the cause within itself."* The sense in which this was taken is, that some central power or influence in each individual, presided over the formation, nutrition, and vital action of all parts, and correlated them into an harmonious whole, and, in fact, furnished vital influence or power to the separate parts. This is just what the vital principle was assumed to effect in olden times, and, in fact, to ascribe a power of this nature to any, even material parts, such as a central nervous system, under the name of "vital force and power," or "directing agency," or "directing power," is nothing better than the old vital principle with a new name.

We see, thus, in the above-mentioned work, the author, then the highest authority in Germany, and, at the same time, an original observer, has, as it were, his face still directed backwards to the old theory of a spirit, or, at least, central power of some kind animating each living individual, and, with the help of the material organs, performing the functions of life. Such a work soon belongs to the past.

In the same year appeared another work on physi-

* Stricker, Syd. Soc., vol. xlvii. p. 1.

ology by an author whose face was directed to the future, and who, abandoning all the ancient theories of spiritual essences as the efficient causes of vital phenomena, referred these latter solely to the inherent properties of the elementary parts, and thus placed himself in harmony with the philosophy of the future.

In this year, 1835, in his masterly work, "Rudiments of Physiology," Dr. Fletcher, of Edinburgh, systematically reviewed, for the last time, the old hypothesis of a vital spirit, or essence, or principle, as the cause of life, and gave it, we may suppose, the *coup de grâce*, for the question is seldom argued now in physiological works, and it is the fashion to treat it as an exploded theory, even by those who have not clearly apprehended the alternative, and are really still following it under other names. That alternative was, however, clearly apprehended by Fletcher, and with such force that he was impelled by the mere course of consistent reasoning to frame an hypothesis of the anatomical nature of the living matter which anticipates, in a remarkable manner, the discovery of the protoplasmic theory of life, which is our subject here. The two chief points laid down by him are—

1. That if vitality do not reside in a separate principle, but depends upon the mode of combination of the elements of the organic parts themselves, there can be no central vital influence communicable to the parts and dominating them, for the vitality of each must be inherent in itself, and, as a property of the material compound, cannot be transferred to the smallest distance; each part, organ, and even cell, there-

fore, possesses a quasi-independent life, and they are all bound together to form an individual merely by the ties of a central nervous system and common circulation or some similar means when these are not present. This is not taught as anything original, and it was a view more or less distinctly expressed by the older physiologists, *e.g.*, Fallopius.

2. That the property of vitality does not reside equally in the various organic structures requiring such different physical properties, but is restricted solely to a universally-diffused, pulpy, structureless matter, similar to that of the ganglionic nerves and to the gray matter of the cerebro-spinal nervous system. This is, as far as I am aware, a perfectly original hypothesis.

No doubt it is easy enough to perceive that the invocation of a spiritual principle which shall cause common, chemically-combined matter to display the powers of life, as an explanation, is no more philosophical than to believe in the capacity of such agencies as witchcraft and magic to do work without adequate physical power. But when we come to particulars, and ask the physiologist who asserts that vitality is merely the property of a certain chemical combination of matter just as fluidity aquosity is the property of the chemical compound we call water, how it comes that from no known chemical action or process can we obtain results the least like living action, he is at fault. He can name and define by chemical tests the matter after death, which a moment before was living, but he cannot now perform a single vital function with this very matter presumed to be identical, and if he does

not fall back upon the notion of some essence or principle which escaped at death, leaving the chemical
compound the same, he speaks with extreme vague-
ness of a peculiar variety of force called vital, evolved
from the chemical force of the food by subtle chemical
processes which we cannot as yet imitate in the laboratory. It is easy to show that no possible form of
force, as defined in physics, could compel chemical compounds, such as albumen, fibrin, and others presumedly
making up the living matter, to act differently from
the manner in which they must act according to their
molecular composition, as albumen, protein, and the
rest.

The difficulty was felt and acknowledged by
Fletcher, and instead of evading it, he met it at once,
and declared that the truly living matter was not in
simply a somewhat different chemical state from that
in which it exists after death—such a statement would
be a mere bald truism—but that the elements are in a
state of combination not to be called chemical at all in
the ordinary sense, but one which is utterly *sui generis*.
That, in fact, no albumen, fibrin, myosin, protagon, or fats
exist at all in the living matter, but that the sum of
the elements of all these is united into a compound,
for which we have no chemical name, and of the complex mode in which the atoms are combined we can
form no idea; and it is only at the moment of death
that those chemical compounds, with which we are
familiar, take their origin. In fact, that death means
simply the resolution of this complex combination
into the simpler compounds, albumen, fibrin, and the
rest, which we find on analysis. Among the expres-

sions of Fletcher on this point, I may quote the following :

"It is only at the instant of the cessation of the vitality of each organized tissue that these compounds or reputed proximate principles are formed—at that instant when the power called chemical affinity succeeds another power which may be called vital affinity, and by which it had been previously superseded, and common chemical compounds are all that is left of that organized mass into which the elements had been associated." Nor is this "power called vital affinity" any essence or force added to the living matter, for "irritability or vitality is a property of organized or living matter, as characteristic of this as inflammability of phosphorus, or elasticity is of ivory." Again : "The process of secretion, by which the ultimate ingredients of all vegetable and animal compounds, whatever they may be, are brought together, is perhaps an infinitely more subtle and searching power than that of common chemical affinity." Nevertheless, it is not anything foreign to the properties of matter, for he adds, "Secretion is a process, although not identical with, still analogous to, common chemical affinity."*

With such sharply defined distinctions between the chemical and vital state of combination of the atoms

---

* The clear expressions of Fletcher gave no countenance to any ambiguity respecting the nature of his living or "irritable matter," such as has lately been experienced respecting protoplasm by the application of that word both to living and dead matter. "Chemical analysis accordingly must be considered as useful in showing us, not what such matter *was* composed of while it possessed vitality, but what it *is* composed of afterwards" (135). And, he adds in a note, "The grave-digger, in Hamlet, spoke more 'by the rule' in these matters—'One that was a woman,' says he, ' but, rest her soul, she is dead.'"

of matter, as might be expected Fletcher gives no
countenance to the idea of any intermediate stage
between them—any stage, as it were, common to both
which would permit the gradation of one into the
other. There is no such thing as vito-chemical in the
sense of partaking of both states.   On the contrary,
the division is sharp, abrupt, and absolute, and between them is an unfathomable gulf.

Vitality is thus a property inherent in each particle
of the living matter, and as all the parts of a complex
organism differ in function, each part has a specific
kind of vitality peculiar to itself.   An individual of
any species is thus a complex congeries of a number of
subordinate quasi-independent living units, whose life
is complete in themselves.   It is impossible even to
touch upon the large question of the development of
the germ into the harmonious arrangement of different
organs and parts in perfect adaptation to their purpose, but it may be stated that in the absence of any
central, overruling, semi-rational, vital principle, Fletcher
holds "that the development of those parts is immediately effected by certain inherent powers, of a different
nature indeed, but not less definite in their operation
than those which determine the crystalization of a
mineral " (i. 65).

With respect to the second proposition, that this peculiar property of vitality does not reside in the tissues
indiscriminately, but in one anatomical element alone,
it is sufficiently obvious that as the various tissues
differ extremely in their physical properties, and these
latter are almost exactly the same after as before
death, it is hardly to be expected that the living

matter can die into or rearrange itself in a short time
into a number of different forms, which shall possess
exactly the same physical properties in the vital as in
the ordinary state of combination. It is likewise to
be expected that as the vital or metabolic molecular
changes in the living matter must be very rapid and
complicated, the physical state of it cannot be hard or
rigid, and this agrees with what has been long known of
the parts in which life is most active and intense, viz.,
the gray matter of the nervous system. Moreover,
we know that there is no example of life existing in
any gaseous or purely liquid fluid. These considerations narrow the probable field of the seat of vitality
very much, and the following question thereupon
is raised by Fletcher :—" Admitting that irritability or
vitality, general and specific, is a property of the
organized solids alone, it becomes a question of the
highest interest whether it be directly inherent in
each of the organized tissues, either of plants or
animals, or whether it merely appears to be possessed
by them all in virtue of some one which is universally
distributed over the organized being, and inextricably
interwoven with every other " (ii. p. 55). He ex-
amines this question, and by an interesting train of
reasoning, based chiefly on arguments derived from
comparative anatomy and physiology, he comes to
the conclusion that vitality is not inherent in any
liquid, nor in any of the rigid structures, and
that it is only in virtue of a specially living matter,
universally diffused and intimately interwoven with
its texture, that any tissue or part possesses vitality.
Therefore, he " must deny any direct participation in

irritability or vitality to those peculiar aggregations of
matter which go to form respectively the cellular,
dermoid, mucous, serous, vascular, fibrous, osseous, car-
tilaginous, or muscular tissues," and also to the white
matter of the nerves. Thus every one of the structures possessing any degree of rigidity, usually de-
nominated the living tissues, is in reality dead just
as much as cuticle, hair, nails, and all the pure fluids.
The only truly living matter consists of the gray
matter of the ganglionic nerves, which he held to be
universally diffused, and the gray matter of the brain
and spinal marrow.

The physical and chemical description of this one
true and only living matter is that of "a pulpy, translucent, homogeneous matter, yielding, after death,
fibrin." Thus we have the remarkable conclusion
that all that is properly called structure and gives
form and beauty and fitness for purpose to animals
and plants is dead, and composed of merely chemically
combined elements, just as we find it after death.
Here then is an ample field for the display of those
mechanical and chemical actions, which are certainly
largely represented in the functions of living beings,
without trenching on the truly vital actions. We may,
without difficulty, now perceive how the bones give
firmness and support; how the teeth grind the hardest
substances; how the arteries and veins form a perfect
system of conduits for nutrient fluid; how the fibrous,
elastic, and connective tissues perform their respective
physical functions; how the muscles form an apparatus, admirably adapted for the physical conditions of
motion in a particular direction, while a purely vital

process may be concerned in the perception of the
stimulus and transformation of the needful force ; how
osmosis, chemical fermentation, interchange of oxygen
and carbonic acid by the hæmoglobin, and all the
various processes, strictly chemical and physical of
animals and plants, are performed in harmony with
vital actions, properly so-called. These last, residing
in this " nitrogenous, pulpy, translucent, homogeneous
matter, yielding, after death, fibrin," and which is
everywhere interwoven with the tissues according to
the degree to which they can be called living tissues,
of course, must vary in strict dependence on the
changes in quantity and quality of this marvellous
combination of matter—so utterly unlike ordinary chemical compounds, and which alone possesses the faculty
of growth or self-renewal and increase from heterogeneous matter. With every vital action, including formation and absorption of tissue and secretions, assimilation, respiration, generation ; with every evolution
of force ; with every sensation, thought, and act of
volition, some portion of this wonderful substance
must pass from the vital down to the chemical state—
must be consumed, in fact—and a corresponding quantity of new living matter assimilated from the pabulum.

The process of assimilation, he held, was always truly
vital, and the components of the tissues were never
absorbed and merely deposited unchanged from the
nutrient fluid, but were always, however near in composition, decomposed, and their elements rearranged in the
process ; and so marvellous is the power of the living

matter in analysis and in producing new syntheses, that he thinks reasons are not wanting to lead us to suppose that it may resolve and transmute the so-called simple elements themselves.

The intimate dependence of every phase of vital action, and of the different properties of the living matter, throughout the whole range of animals and plants, on a corresponding change of molecular composition of the living matter, made up as it is of the same few ultimate elements, cannot as yet be demonstrated experimentally, because finer analyses of the products of different varieties of living matter, after death, are still wanting owing to the difficulty of its isolation ; and also, because this, no doubt, may consist in the mere arrangement of the atoms in the extraordinarily complex molecules of the living matter, as we see in ordinary chemistry with the various series of isomeric bodies.

Not only is every vital action traced to molecular change and to consumption and regeneration of this structureless, semi-fluid matter, combined in a way entirely *sui generis,* but the initiation of these changes is brought by Fletcher into absolute dependence on stimuli, and all spontaneity or autonomy is denied to matter in the living just as in the dead state. Thus every physiological action is reduced to dependence on adequate causes exactly in the same way as the phenomena of the inorganic world. The necessity for stimuli to all muscular motion, and to the senses, and to many secretions, is generally recognized, but that they are equally essential for growth, development, and nutrition, is overlooked by many physiologists, who still speak of certain actions and functions as spontaneous, and as thereby manifesting a distinction between the organic and inorganic kingdoms of nature.

CHAPTER II.

CELL THEORY BEFORE 1860.

THE progress of physiological knowledge from the time of Fletcher may be said to be bound up in the history of the cellular theory, which may be considered practically to have begun in 1838, when the microscope was sufficiently perfected to give a solid basis for the observation of facts. The hypothetical anticipation of it by various authors in preceding times, although interesting, need not detain us, and I may merely refer those desirous of studying it to Professor James Tyson's excellent work on the cell doctrine. Taking up the subject from 1838, I will endeavour to select from the bewildering mass of details and conflicting statements which have accumulated since then those points which have a definite bearing on the principles of the question. This may be best done by tracing the cell doctrine in its complete form up to the time when the cell was generally accepted as the ultimate elementary unit of life; then, by tracing again from the beginning the doctrine which is believed by many to have now supplanted it, viz., that the place of the cell is to be taken by one of its constituents—the protoplasm.

This may be given up to the period of its full development by Dr. Lionel Beale, and after that a review of the present state of knowledge upon these, in some respects, rival theories. This method will, I think, conduce to clearness of understanding the subject, better than the strictly chronological method in which both are mingled together.

Schleiden, who was the founder of the cell theory, though by him restricted to plants, defines the vegetable cell as " the elementary organ which constitutes the sole essential form-element of all plants, and without which a plant cannot exist ; and as consisting, when fully developed, of a cell wall composed of cellulose, lined with a semi-fluid, nitrogenous coating." With him, therefore, the cell consisted of two parts, viz., a vesicle and semi-fluid contents. In plants the cell forms are distinct, and easily recognized, and thus, when the conception of a similar elementary organ was extended to the animal kingdom by Schwann, in 1838, it is not to be wondered at that the cellular form was expected to be universal. Schwann added to Schleiden's two elements a third— the nucleus—which he deemed also of essential importance, and to be present in all cells, if not always, at least in some stage of their existence. On his authority this threefold doctrine of the cell became universally prevalent for a time. I give here Schwann's original definition of his theory, as some points in it have been overlooked or forgotten in the mass of controversial writing this subject has provoked :—

" The following admits of universal application to the formation of cells :—There is, in the first instance, a *structureless* substance present, which is sometimes quite fluid, at others more or less gelatinous. This substance possesses within itself, in a greater or lesser measure, according to its chemical qualities, and the degree of its vitality, a capacity to occasion the production of cells. When this takes place, the nucleus usually appears to be formed first, and then the cell around it. The formation of cells bears the same relation to organic nature that crystallization does to inorganic. The cell, when once formed, continues to grow by its own individual powers, but is, at the same time, directed by the influence of the entire organism in such manner as the design of the whole requires. This is the fundamental phenomenon of all animal and vegetable vegetation. It is alike equally consistent with those instances in which young cells are formed within parent cells, as with those in which the formation goes on outside of them. The generation of the cells takes place in a fluid, or in a structureless substance in both cases. We will name this substance in which the cells are formed, cell-germinating material (zellenkeimstoff), or cytoblastema. It may be figuratively compared to the mother-lye from which crystals are deposited " (Syd. Soc., 1847, p. 39).

We perceive that Schwann added little to the conception of Schleiden, but he extended it to all organisms, whether animal or vegetable, and applied it with considerable success to the details of the formation of animal tissues, in which process the whole three cell elements were assumed to play a part, and a distinctively vital one. Schwann was also more decided in admitting the free origin of cells in a blastema than Schleiden.

However, in proportion as the cell theory was applied more extensively in the animal kingdom it became more and more difficult to maintain the threefold nature of the cell.

In giving now a general view of the development of the cell theory I will not attempt to give a complete history, apportioning to each observer his share of the merits in the building up of the theory. I will merely quote in full, or analyze, those memoirs which mark the chief stages in its progress, with a few connecting observations.

Several important changes were introduced into the above theory before it became for a time established, in spite of these changes, still as a cell theory.   In 1841 Henle adopted the cell theory of Schleiden and Schwann, but pointed out the multiplication of cells by division and budding.   In the same year Dr. Martin Barry showed the reproduction of cells by division of the parent nucleus, and confirmed Schleiden in the importance of the nuclei, as new cell-formers.   But the first important contributions to the cell theory, after Schwann, were the memoirs of J. Goodsir, in 1842 and 1846 ("Anatomical Memoirs," vol. ii.), and they still remain probably the most important till the time of Dr. Beale.   The first was on secreting structures, and as growth and secretion are substantially the same vital processes the theory of Schwann received elucidation and development from another side, as it were.   Since the time of Malpighi, the secreting glands were known to be composed essentially of tubes with blind extremities, but the exact seat of the vital process of secretion was not agreed upon. By Fletcher, and probably the majority of physiologists, it was supposed to be the walls of the capillary vessels.   Schwann suggested it was in the epithelium of the mucous membrane of the ducts, and Purkinje hypothetically placed it more definitely in the nucleated epithelium, but did not verify that hypothesis by observation.   Goodsir brings together a number and variety of observations on the secreting organs of animals from the mollusca up to mammals, and finds a common character running through them all, viz., that the specific secretion is found inside the nucleated epithelial cells, between the nucleus and the wall of the cell.   The animals were selected on account of the striking colour possessed by the secretion, such as the *Loligo sagittata,* on account of its ink-bag ; the *Phallusia vulgaris,* for the dark-brown fluid of its hepatic organ ; the *Janthina fragilis,* for the purple fluid secreted by the inner surface of its mantle, and which is the source of the Tyrian dye, &c.   In many cells the secretion is so transparent and colourless, that ocular proof of its formation within the cell is impossible, and no chemical test could be applied. In the first publication of this memoir, in the " Transactions of the Royal Society of Edinburgh," he follows mainly Schwann, thinking that the nucleus is the reproductive organ of the cell, and has nothing to do with the formation of the secretion. He adds—" I believe that the cell wall itself is the structure by the organic action of which each cell becomes distended with its peculiar secretion at the expense of the ordinary nutritive medium which surrounds it" (p. 417). But in the republication of the article, in 1845, he says—" The ultimate secreting structure is the primitive cell endowed with a peculiar organic agency, according to the secretion it is destined to produce. I shall henceforward name it the primary secreting cell. It consists, like other primitive cells, of three parts—the nucleus, the cell wall, and the cavity. . . . The secretion within a primitive cell is always situated between the nucleus and the cell wall, and *would appear to be a product of the nucleus* " (p. 417).

He then states at p. 426 :—" Since the publication of my paper on the secreting structures, in the ' Transactions of the Royal Society of Edinburgh,' in 1842, I have satisfied myself that I was in error in attributing to the cell wall the important function of separating and preparing the secretion contained in the cell cavity. The nucleus is the part which effects this. The secretion contained in the cavity of the cell appears to be the product of the solution of successive developments of the nucleus, which in some instances contains in its component vesicles the peculiar secretion, as in the bile cells of certain mollusca, and in others becomes developed into the secretion itself, as in seminal cells. In every instance the nucleus is directed towards the source of nutritive matter, the cell wall is opposed to the cavity into which the secretion is cast. This accords with that most important observation of Dr. Martin Barry, on the function of the nucleus in cellular development."

Having, as above, described the nucleus as the generative, or reproductive organ of the cell, he now shows that reproduction and secretion are in reality varieties of the same process. "There are, in fact," he says, "three orders of secretions

2

—1. A true secretion—*i.e.*, matter formed in the primary secreting cell cavities ; 2. A mixture of fluid formed in these cell cavities, with the developed or undeveloped nuclei of the cells themselves ; and, 3. It may be a number of secondary cells passing out entire." These he supports by observations on the testicles of the *Squalus cornubicus*, which show a continual production of cells within cells, which become developed into complete spermatozoa and are thrown off, the glandular parenchyma being in a constant state of change, contemporaneous with and proportioned to the rapidity of the secretion ; therefore " there are not, as has hitherto been supposed, two vital processes going on at the same time in the gland, growth and secretion, but only one, viz., growth—the only difference between this kind of growth and that which occurs in other organs being, that a portion of the product is from the anatomical condition of the part thrown out of the system" (p. 422).

In 1845 he adds the following :—

" I have also had an opportunity of verifying, and to an extent which I did not at the time fully anticipate, the remarkable vital properties of the third order of secretion, referred to in the memoir to which I have just alluded. The distinctive character of secretions of the third order is, that when thrown into the cavity of the gland, they consist of entire cells, instead of being the result of the partial or entire dissolution of the secreting cells. It is the most remarkable peculiarity of this order of secretions that after the secreting cells have been separated from the gland, and cast into the duct or cavity, and therefore no longer a component part of the organism, they retain so much individuality of life, as to proceed in their development to a greater or less extent in their course along the canal or duct, before they arrive at their full extent of elimination. The most remarkable instance of this peculiarity of secretions of this order is that discovered by my brother, and recorded by him in a succeeding chapter. He has observed that the seminal secretion of the decapodous crustaceans undergoes successive developments in its progress down the duct of the testis, but that it only becomes developed into spermatozoa after coition, and in the spermatheca of the female. He has also ascertained that, apparently for the nourishment of the component cells of a secretion of this kind, a quantity of albuminous matter floats among them, by absorbing which they derive materials for development after separation from the walls of the gland. This albuminous matter he compares to the substance which, according to Dr. Martin Barry's researches, results from the solution of certain cells of a brood, and affords nourishment to their survivors. It is one of other instances in which cells do not derive their nourishment from the blood, but from parts in their neighbourhood, which have undergone solution ; and it involves a principle which serves to explain many processes in health and disease, some of which have been referred to in other parts of this work."

From these it appears obvious that much of what was only plainly understood long afterwards is anticipated, particularly that the vital action resides in one constituent of the cell, chiefly if not entirely, here called the nucleus, and also the independent life of detached and migratory particles of living matter.

The paper on "Centres of Nutrition," 1845, does not bear so much on our subject, except in the unqualified assertion that cells never arise except from pre-existing cells, which was afterwards adopted by Remak and Virchow. It contains, probably, all that is true in the theory of cell territories which Virchow puts forth without sufficient acknowledgment of Goodsir's priority, and does not contain the addendum of Virchow, viz., the juice canals which Beale has shown to have no existence.

The comparative unimportance of the cell wall was also shown by Naegeli in 1845, and by Alexander Braun in 1851, who both, in fact, maintained that it was non-essential. The credit is usually, however, given to Leydig, of having, in 1857, first decidedly declared that the cell membrane was non-essential, and that the cell consisted of " a soft substance enclosing a nucleus." This was confirmed in 1861 by Max Schultze, who observed that very many of the most important kinds of cells were destitute of membrane, and he defined the cell as " a little mass of protoplasm, inside of which lies a nucleus. The nucleus, as well as the protoplasm, are products by partition of similar

components of another cell." The cell wall being given up, the
threefold nature of the cell as the elementary vital unit disappears, and physiologists go back to Schleiden's idea of its dual
nature ; but the two elements are now nucleus and cell contents, and as long as the cell theory is maintained, the importance of the nucleus becomes essential. " A plasma-lump
without a nucleus is no longer a cell," says Häckel ("Gen.
Morph.," i. 273). Upon the consistence, and structure, and
physiological nature of the nucleus, the most conflicting and
manifold statements and opinions are given by botanists and
zoologists. It would be tedious and superfluous to go into
these in detail, so I will merely note here the position it occupied in the cell theory up to 1860. It is said to be always
round, or a more or less prolonged oval, whatever be the shape
of the cell, and to be of the same chemical composition as the
protoplasm in which it lies embedded, or hardly distinguishable from it. It frequently contains within it a smaller similar
body, the nucleolus, and sometimes within that may be detected
a still smaller one, the nucleolinus. The essentiality of this
body, and its supposed functions in the cell theory, were thus
summed up by Virchow, in 1858 :—

" The nucleus plays an extremely important part within the
cell . . . . less connected with the function and specific office
of the cell, than with its maintenance and multiplication as a
living part. The specific (in a narrower sense, animal) function
is most distinctly manifested in muscles, nerves, and gland cells ;
the peculiar actions of which—contraction, sensation, and secretion—appear to be connected in no direct manner with the
nuclei. But that whilst fulfilling all its functions the element
remains an element, that it is not annihilated nor destroyed by
its continual activity—this seems essentially to depend upon
the action of the nucleus " ("Cellular Path.," p. 10). Doubts,
however, were thrown before this on the universality of the
nucleus, and in 1854, Max Schultze had described a non-nucleated Amœba found in the Adriatic—the Amœba porrecta. The
phenomena spoken of by Virchow will afterwards be seen to
be capable of quite different explanation, nevertheless, the cell
theory in the altered form (contents and nucleus) above noted

was accepted for many years in medicine ; the cell being in the words of Virchow "the ultimate morphological unit in which there is any manifestation of life." The development of cells form a free blastema was again favoured by Todd and Bowman in 1856, but since the adoption of Goodsir's proposition by Virchow, it has been finally abandoned, and the aphorism of that author *omnis cellula e cellulâ* has been substantially accepted by all. It must be borne in mind that while the non-essentiality of the cell wall to the completeness of the cell was generally accepted, nevertheless, as yet, no one had distinctly denied to it in all cases the participation in truly vital functions when present, and it is especially to the cell membrane that Schwann attributes the power he first named metabolic, and which it is here proposed to accept as synonymous with vital.

By the cell theory, we have thus arrived at a system by which "every animal presents itself as a sum of vital unities, any one of which manifests all the characteristics of life " (Virchow, p. 13). Likewise the specific characters of the life of each part is inherent in these unities themselves, and is not assigned to them by any central life or power of a spiritual or other nature. Nor can anything of the nature of life be communicated from one of these unities to another except by way of growth and subdivision. By the general acceptance of this theory the first of the principles contended for by Fletcher is thus seen to be established, and the hypothesis of a single central vital principle, or *anima*, or spirit which gives the unity and vital character to individuals in the animal kingdom is shown to be superfluous and inconsistent with the facts. As observed by Schwann, "the whole organism subsists only by means of the reciprocal action of the single elementary parts "—the

expression reciprocal action, being taken in its widest
sense as implying the preparation of material by one
elementary part, which another requires for its own
nutrition. Thus the majority of the individual cells
may be unable to subsist when separated from the
whole organism, because it is only while together
they can obtain the nutriment and other conditions
requisite for continued life. Therefore, "the cause of
nutrition and growth resides, not in the organism
as a whole, but in the separate elementary parts—
the cells. The failure of growth in the case of any
particular cell, when separated from an organized body,
is as slight an objection to this theory, as it is an
objection against the independent vitality of a bee,
that it cannot continue long in existence after being
separated from its swarm. The manifestation of the
power which resides in the cell depends upon conditions to which it is subject only when in connection
with the whole (organism) " (Syd. Soc., 1847, p. 192).

Nevertheless when we see that in the cell theory
proper, the smallest living unit is a compound possessing structure, viz., the wall and contents which are
differently constituted both physically and chemically,
we can no longer conceive that its vitality can be the
property of the matter of which it is composed. For
a property must be present in full measure and inseparable from the smallest indivisible molecule of
which any mass is composed. For instance, no one
would think of attributing aquosity or the distinguishing properties of water to it only in the form of single
liquid drops or single crystals of ice, far less to a drop
surrounded by a coating of something else, but always

to the smallest indivisible molecule of hydric oxide in whatever physical state.   Therefore we are again thrown back upon the notion of a concrete life added to a certain compound organized structure, a notion not more conceivable or more tenable for simple cells than for the whole individual.   Hence if vitality is to be a property of matter at all, it must be of a physically homogeneous substance, every molecule of which must possess that attribute.   Moreover, in all modifications of the cell theory, even those which allow for the occasional absence of the cell wall, that part when present is believed to take an active part in the strictly vital process of transformation into tissue. Now the cell wall is in all cases solid and possessing a certain rigidity, and in different cells it passes by insensible shades through an infinite variety of degrees of hardness, and of states, many of which are known to be incompatible with life.   Likewise it is continuous with, and shades off into, the intercellular substances which offer an infinite variety of composition, many of them being non-nitrogenous—a composition which we know is incompatible with life; further, the cell wall, or what corresponds to it functionally, passes in other cases by insensible degrees into an infinite variety of true fluid secretions which are soluble and diffusible—which no living thing is.

For these reasons, and that given at p. 7, we are driven to the conclusion that the attribute of vitality cannot reside in anything of the nature of a cell wall, and therefore of a cell taken as a whole.

CHAPTER III.

THE PROTOPLASMIC THEORY BEFORE 1860.

Even before the supremacy of the cell was shaken, biologists began to notice that the cell theory most in vogue through Virchow's compendium was not the theory of Schwann in its complete form, and the opinion was expressed by Max Schultze * that "in many points we must go back to the purer form of the doctrine."† Now what is the pure form? At p. 165 of Schwann's work we read "that in the fundamental phenomena attending the exertion of productive power in organic nature *a structureless substance* is present in the first instance, either around or *in the interior of cells already existing;* and cells are formed in it in accordance with certain laws, which cells become developed in various ways into the elementary parts of organisms."

In respect to this cytoblastema, or amorphous

---

* "Protoplasma der Rhizopoden," p. 63.
† Perhaps we should go further back to C. F. Wolff, who traced back the point of departure of all development, both animal and vegetable, to a "clear, viscous, solidescible nutritive fluid possessed of no organization."

substance, in which new cells are to be formed, he states that it may exist within ready-formed cells, and outside of them, before they are formed; that it is the matrix and also the nourishment of cells; that it draws its own nourishment from the blood, but that it differs chemically in different parts from that blood, although how that difference is produced he does not explain, for it is afterwards referred to the metabolic power of the future cell wall and nucleus. Moreover, the cytoblastema is said to possess vitality in various degrees, but that may also be totally destitute of it, as, for example, a boiled infusion of malt is said to be the cytoblastema of yeast cells. Now, if we subtract from these opinions what relates to the implied origin of living cells from dead chemical matters, which part has been eliminated by nearly all men of science, we find that the primary formative matter is a structure-less substance, possessing vitality, and proceeding from pre-existing living matter. And it will not be difficult to explain the confusion of the above statements by the simple supposition that portions of it (even if too minute to escape detection with the then existing microscope) were present or not in the different circumstances: and we may thus reconcile the observations of this great pioneer with the discoveries of future investigators. Let us see how far this is borne out by the subsequent history of this structureless living matter.

The first notice of what appears to be identical with the structureless living matter of Schwann, and the irritable or living matter of Fletcher, is found in a memoir by Dujardin, in vol. iii. of the "Annales des Sciences Naturelles," in

All Doom

1835. When speaking of the Rhizopoda, which he thus names for the first time, he says, "On ne peut voir là de véritable tentacules, c'est une substance animale primaire qui s'etend et pousse en quelque sorte, comme des racines." And he speaks of the simplicity of the tissue, calling it a "sorte de mucus doué du mouvement spontané et de la contractilité" (p. 314). His next mention of the subject is in vols. iv. and v. of the same work. In this paper which treats of certain Rhizopoda, chiefly the Gromia oviformis, and Miliola and Amœbea, he shows the absence of any investing membrane. He first uses the term sarcode in describing the movements of the *Proteus tenax*, stating that, before its death, "il se montre entouré de cette matière diaphane glutineuse que j'appelerai *sarcode*, et qui exsude à travers le sac membraneux" (p. 354).

In another lively individual he notices and figures two globular exudations of sarcode, which changed their place during the movements of the animal, and his subsequent descriptions and plates give those movements and bulgings which all who have observed Amœbæ and similar organisms, are familiar with. He attributes the movements to an inherent force in the mass of the sarcode, but that it is rather a force of extension than contraction which enables the Rhizopoda to push out these prolongations (359).

He then enters on a treatise in detail (364) upon the sarcode. "I propose to name thus, what others have called a living jelly, viz., that glutinous substance, diaphanous, insoluble in water, contracting into globular masses, sticking to the dissection needles, and thus capable of being drawn out into thread like mucus, and, finally, which is found in all the inferior animals interposed between the other structural elements" (367). He then describes its contractile movements, and thus accounts for the formation of vacuoles, which had been erroneously taken for stomachs by Ehrenberg. The rest of the memoir is devoted to the description of various flagellate infusoria, and to demonstrating the absence of those organs which had been given to them by Ehrenberg.

The next in chronological order, viz., 1838, were the views of Schleiden, which are peculiarly interesting as prefiguring the conclusions ultimately come to by Dr. Beale. Schleiden was struck by the observation by Robert Brown, in 1833, of the frequent presence of an opaque spot in the cells of the epidermis of the orchideæ, which he named the nucleus, but did not follow out the matter further. Schleiden, on the other hand, finding it constantly present in the cells of young embryoes, and in the newly-formed albumen, perceived its significance in the development of the cell, and, finally concluding that it was an universal elementary organ of vegetables, and that by which all cells were formed, named it the cytoblast, or cell bud. The colour of this is yellowish, or white, or at times so transparent that in some plants, *e.g.*, the helvelloids, it is scarcely perceptible from that cause. He observed, also, what had escaped the notice of Robert Brown, that some contain one or more circular bodies which correspond to what have been since called the nucleoli, and that these were formed earlier than the nuclei, or cytoblast, and can develop into them, and hence into cells.

In his first description of the formation of cytoblasts he is not very clear. He states that in the gummy matter from which the tissues of plants are formed a number of granules make their appearance ; and in this mass organization takes place, and a gelatinous matter is formed which is ultimately converted into cellular membrane and fibres. In the above gummy mass, after the granules, cytoblasts make their appearance, and when full-sized they form cells, as described thus at p. 238 : "A delicate transparent vesicle rises upon the surface. This is the young cell, which at first represents a very flat segment of a sphere, the plane side of which is formed by the cytoblast." The vesicle gradually expands and increases beyond the margin of the cytoblast, and quickly becomes so large that the latter at last merely appears as a small body enclosed in one of the side walls between the laminæ. "In this situation it passes through the entire vital process of the cell which it has formed" if it be not dissolved and absorbed as a useless member. As a general rule, this last happens ; but there are exceptions in which it remains, such as the *orchideæ* and *cacteæ*, which continue through life in a lower stage of development, and in some other plants, and also in pollen granules. In all this is prefigured the more modern doctrine which makes the nucleus, here alluded to, the truly vital part : but that is not distinctly recognized by Schleiden, although, in the remark which immediately follows, he comes very near to the general principle : "Lastly, many hairs, particularly such as exhibit motions of the sap within their cells, retain the cytoblasts. It is at the same time remarkable, and a proof of the close relationship which the cytoblast bears to the whole vital activity of the cell, that the little currents, which frequently cover the entire wall like a network, always proceed from and return to it, and that when *in statu integro* it is never situated without the currents" (p. 240).

From the foregoing account of the origin of cytoblasts in the gummy fluid, Schleiden is generally stated to be an advocate for exogenous free cell-formation, and Tyson (p. 33) says, "that it involves a spontaneous generation of the cell." On an attentive study of the original memoir on Phytogenesis, this seems to me not borne out, although Schleiden is not so clear as might be wished. For, farther on, he repeats an account of the process, stating that it takes place in the gummy matter within the embryonal cell first, and then in the cells descended from that. The origin of the granules remains doubtful, as from their extreme minuteness, and often transparency, they cannot at first be detected ; and frequently a cell will be found to be absorbed, and two new ones appear in its place, without our being able to detect the stages of the process : and as we know that the smallest nucleolus is capable of becoming a cytoblast, "then indeed we are forced to confess that the imagination obtains ample latitude for the explanation, in every case of the generation of infusorial vegetable structures, even without the aid of a *deus ex machinâ* (the *generatio spontanea*)." He traces the growth of the whole plant to a repetition of what takes place in the growth of the embryo, which consists in the formation of cells within cells. "After the first cells, generally few in number, are formed, they rapidly expand to such an extent that they fill the pollen tube, which soon ceases to be perceptible as the original enveloping membrane ; but at the same time, several cytoblasts originate in the interior of each of these cells, and generate new cells, on the rapid expansion of which the parent cells also cease to be visible, and become absorbed. The same process is repeated indefinitely" (p. 253). From all this it is not difficult to see that, without full appreciation of the fact,—his mind being occupied with the idea of cell-form—his observations of nature are in harmony with the doctrine that all these granules, nucleoli, and cytoblasts are masses of living matter directly descended from pre-existing living matter, furnished by the germ, and which not only form but sustain all the vital processes in the formed cells.

The next contribution to the subject is one of particular interest as here for the first time we meet with the word protoplasm. It consists of a memoir by Hugo von Mohl, first published in 1844, but we may give an analysis of his views taken from his later work on the " Vegetable Cell," 1853, p. 36.

" If a tissue composed of young cells be left some time in alcohol, or treated with nitric or muriatic acid, a very thin finely granular membrane becomes detached from the inside of the wall of the cells in the form of a closed vesicle which becomes more or less contracted and consequently removes all the contents of the cell, which are enclosed in this vesicle, from the wall of the cell. Reasons hereafter to be discussed have led me to call this inner cell the *primordial utricle*. Iodine colours it yellow, and it is therefore probably always nitrogenous. Cellulose cannot be found in it, and the compound of which it is composed is as yet unknown. The primordial utricle disappears again with the thickening of the walls, of the vessels, the cells of the wood, of the pith, of the inner part of the petioli, and of thick leaves. . . . In the centre of the young cell, with rare exceptions, lies the so-called nucleus of Robert Brown. . . . . The remainder of the cell is more or less densely filled with an opaque viscid fluid of a white colour having granules intermingled in it, which fluid I call *protoplasm*. This fluid is coloured yellow by iodine, coagulated by alcohol and acids, and contains albumen in abundance, whence young organs are always very rich in nitrogen. . . . During the growth of the cell, irregularly scattered cavities are formed in the protoplasm ;

these are originally isolated, and very frequently present
a most deceptive resemblance to delicate-walled cells;
subsequently, however, they become blended together in
many directions; the protoplasm is then accumulated at
one side in the vicinity of the nucleus; on the other side it
coats the inside of the primordial utricle, and these two collections are connected together by thread-like processes,
which are sometimes simple and sometimes branched, so
that the nucleus appears as if suspended in a spider's web in
the centre of the cell. An internal movement of the protoplasm
now begins to be visible—originally no definite arrangement
can be perceived in it; but the more the protoplasm changes
from the uniform mass which it originally formed into the condition of threads, the more distinctly it can be seen that each
of these threads represents a thinner or thicker stream, which
in one thread flows from the nucleus to the periphery, turns
round there and flows back again in another thread. The
thickness, the position, and the number of these threads are
subject to constant change, which shows beyond a doubt that
the currents move freely through the watery cell sap—and are
not enclosed in membranous canals. . . . . The nucleus retains
its central position in many cases even when the cell is fully
developed; e.g., in *Zygnema,* but it usually becomes gradually
withdrawn towards one side of the wall of the cell where it
becomes attached by its viscid investment to the primordial
utricle, but always forms the centre of the currents of sap.
The circulation of the protoplasm is very slow. I determined
it in the hairs of the filaments of *Tradescentia* on an average
of 1-500th of a line per second." He gives the details of the
mode of circulation of the sap in the cells which are familiar
to all. After noticing that the nucleus and the protoplasm
gradually diminish with the age of the cell and the circulation
stops, he gives exceptions in which it is retained by the full-
grown cell, e.g., the stinging hairs of the nettle, the hairs of
curcubitaceous plants and of the filaments of Tradescentia, &c.
In those, active vital functions persist and we see the protoplasm remains in its integrity. Speaking of the apparently
cell-like appearance of the vacuoles in the protoplasm he indi-

cates its physical state as follows : " Yet if we reflect that the protoplasm is a viscid fluid which, as its delicate currents show most distinctly, does not mix with the watery cell sap this appearance becomes comprehensible enough ; the protoplasm bears the same relation to the cell-sap as a frothing fluid does to the air contained in its bubbles. The unceasing flow and continued transformation of the mass of the protoplasm furnish most distinct proof that we have to do with a fluid and not with an organized structure."

The next important contribution to the subject was made by Naegeli * in 1844 and 1846. Although he admits the possibility of spontaneous generation of the first cells of plants and animals apparently for the sake of theoretical completeness, yet he lays down the law that all ordinary cell formation takes place exclusively from pre-existing cells. This applies both to the vegetable and to the reproductive] cell formation (134). He denies Schleiden's theory that mucilage-granules unite to form nucleoli, and these to form nuclei, and says that the embryo-sac when cell formation is commencing contains no granules (105). In all cases of cell multiplication it is the contents which divide, grow, and cover themselves with membrane ; in vegetative cell formation only two cells are formed by the division of the parent cell ; in the reproductive there may be from one to an indefinite number (137). The individualization of the contents for the purpose of cell formation takes place under four forms which he describes. But they all include two stages : "first the isolation or individualization of a portion of the contents of the parent cell ; the second consists in the origin of a membrane around individualized portions of the contents " (123). In all these cases it is the mucilaginous cell contents corresponding to the protoplasm of Mohl which is the active agent, and " whether the mucilage be free or lie in contact with a membrane it makes no difference in its function " (125). " The cell membrane is an investment lying upon the surface of the contents, secreted by the contents themselves " (128). It is never a deposit from the fluid as Schleiden

* " On the Nuclei, Formation, and Growth of Vegetable Cells." Ray Society, 1849, p. 95.

and Schwann say, and does not originate by any chemical action of one kind of substance upon another, but is an organic process—in fact a secretion (128). Likewise "the membrane bears no immediate share in the production of the new cell, it is merely the *contents* of the parent cell that here come into consideration" (138). The young cell first appears as a layer of mucilage surrounding a nucleus. Then a membrane becomes visible on the surface of the mucilage. Then as the cell expands there is a hollow in the interior, and the mucilage remains on the wall as a thin layer. This layer is what Mohl has named the primordial utricle. It is thicker at the place where the nucleus lies, and the latter is often imbedded in it : and it was this circumstance no doubt which made Schleiden think that it lay between the two layers of the cell wall, which cannot be, as Naegeli has seen the nucleus become detached (111). This "mucilaginous layer (primordial utricle) is always present in cells as long as they retain their vitality," and "it secretes organic unazotized molecules which form the new thickening layers" [of the cell wall] (124). "These facts prove that organic unazotized molecules are secreted on the surface of living vegetable mucilage, and these enclose the mucilage in the form of a membranous layer" (124). It is here explained in a note that this "mucilage" corresponds to the "protoplasma" of Mohl. With respect to the nucleus his views are very complicated and not very satisfactory even to himself, but its chief function is the "individualization of the cell contents." In the process of cell multiplication "a nucleus is formed, and this nucleus individualizes a portion of the contents by attraction," and therefore the formation of the new cell takes place as above described. However, in many organisms, *e.g.*, chlorococcus, Hœmatococcus, the germ cells of fungi and lichens, &c., the nucleus may be wanting in the whole process of cell-multiplication (132).

He contests Meyer's view that cells might originate in the homogeneous mixture produced by dissolved matters ; and also that of Mirbel, who said the same of the cambium of the root of the date palm. He also says Endlicher and Meyer are wrong in saying that cells originated in inter-cellular substance. He proves this by showing that in Algæ and fungi (nostoc, palmella, &c.) the gelatinous intercellular mass was produced by the cells—not *vice versâ* (135).

We have thus, in 1846, an anticipation of some very important principles whose application to animal physiology was only recognized long afterwards.

We turn next to the memoir of Alexander Braun, entitled "On the Phenomena of Rejuvenescence in Nature," 1850, translated by the Ray Society, 1850.

Alex. Braun follows Schleiden in holding the cell to be "the simplest sphere of formation in the course of the life and growth of the plant, from which all development starts, which in infinitely varied repetition and modification accompanies the entire development, and to the independent representation of which the conclusion of the development once more returns" (121). He thus describes the cell.

"Examining the individual cell more closely, we must, in the first place, observe that the term *cell* does not correspond exactly to that to which we especially apply it, for we understand as cells, not merely the membranous vesicles or utricles which form the tissue, but also their contents ; we apply the name cell, not alone to the little chamber formed by a completely closed-in wall, within which the vegetable life conceals itself, but also to its living inhabitant, the more or less fluid and inwardly mobile body, which is bounded within the chamber by its more delicate coat (the primordial utricle). The cell is thus a little organism, which forms its covering outside, as the muscles, the snail, or the crab does its shell. The contents enclosed by these envelopes form the essential and original part of the cell ; in fact, must be regarded as a cell, before the covering is acquired. From the contents issues all the physiological activity of the cell, while the membrane is a product deposited outside, a secreted structure, which only passively shares the life, forming the medium of intercourse between the interior and the external world, at once separating and combining the neighbouring cells, affording protection and solidity to the individual cell in connection with the entire tissue. Hence the development of the cell coat, as a product of cellular.

3

activity always stands in inverse proportion to the physiological activity of the cell. In youth, thin, soft, and extensible, the cell coat allows abundant nutrition and advancing growth; subsequently thickened and therewith hardened by the deposit of lamellæ, it compresses the contents within continually narrower boundaries, more and more excludes intercourse with the external world, and puts a term to growth. Thus the life of the plant builds its tomb in the very cell—dies away at last in its own work" (p. 155).

That the origin of the cell precedes its enclosure by a cell membrane is best shown by the cases in which it originates free, that is, without contact with the mother-cell, or where it becomes free by being expelled immediately after its production, e.g., in the swarm cells, or active gonidia of the Algæ. These possess no cell membrane separable from the contents as long as the motion lasts, and must be regarded as bounded merely by the primordial utricle which is intimately connected with the contents (156). Many other proofs of the position are given, that the cell membrane is formed by secretion on the surface as stated by Naegeli, but he is not decided whether the primordial utricle of Mohl exists as a separate envelope, or is a mere lining of protoplasm as stated by Naegeli, and he agrees with the latter in the absence of a nucleus in many unicellular Algæ. In the formation of the starch granules of the hydrodictyon, he considers that the process is by deposition of layers on the surface as was held by Fritsche and Schleiden; and that the starch is like the cell membrane, secreted by the protoplasm. "The idea of origin of cells outside, between or on the surface of existing cells, formerly advocated by Mirbel has proved untenable" (227). "It is a mistake to apply the word cell sometimes to the cell with a membrane, sometimes to the cell without a membrane, and sometimes to the membrane without the cell. Since the contents of the cell constitute the essential part of it, since it forms, before the secretion of the (cellulose) membrane, a separate entity possessing its own proper membranous boundary (the primordial utricle), we must call this internal body the cell proper, unless we restrict the term cell to the enclosing wall or chamber, and give the internal body another name. If the name is restricted to the internal body, we cannot, in the great majority of cases, say that new cells are formed in the old, but merely that they are formed out of the old " (228). He agrees with Mohl, Naegeli, Unger, and Hofmeister, that the division of the cells into two is chiefly, if not universally, the process of development of the tissues of plants (233). He then gives a detailed description of the manifold ways in which this may take place, and then enters on the conjugation of cells, which leads away from our subject.

In 1850 also, F. Cohn makes some important contributions to the general subject in his paper.* Here he recognizes the protoplasm as the contractile element, and as what gives to the zoospore (Schwärmzelle) the faculty of altering its figure without any corresponding change in volume. The protoplasm is said also to possess all the properties both visible in life and to chemical reagents attributed by Dujardin to the *Sarcode* of the infusoria and rhizopods. Therefore, he not only concludes that the protoplasm of botanists is, if not identical, at least in the highest degree analogous to the contractile substance and sarcode of animals, but that this substance—the protoplasm— " must be regarded as the prime seat of almost all vital activity, but especially of all the motile phenomena in the interior " [of the cell] (p. 534). With Cohn, the protoplasm corresponds pretty nearly with the primordial utricle and primordial sac, and primordial cell, which is simply the primordial sac assuming the figure of a cell without any rigid cell membrane (335); when a separation of cell contents takes place he calls the more dense layer of peripheric protoplasm the primordial utricle as Naegeli does (337). But he cannot determine with certainty the presence of any nucleus in this species, and he observes that the protoplasm or primordial sac can sub-divide into a number of segments " without demonstrable influence of a Nucleus " (543). He looks upon the flagella as prolongations of the protoplasm, and therefore protoplasm themselves.

Pringsheim (Untersuchungen über d. Bau u. d. Bildung d. Pflanzenzelle, 1854) maintains that everything that lay within the cell membrane of a living cell might have a complex dis-

* On the Protococcus Pluvialis. Ray Society, 1853, p. 517.

position but consisted essentially of protoplasm and cell fluid. He admits, in the cortical layer of the protoplasm, a distinct arrangement into layers often occurs, but these cannot be differentiated as a membrane—primordial utricle—and subjacent protoplasm. He states that he has coloured blue the outermost layer of protoplasm in *confervæ*, so most probably, he thinks, the primordial utricle is really the most recently formed layer of the permanent cell-wall. In fact, from what appears in the foregoing and what is now generally held, the primordial utricle seems to be merely the outer layer of this protoplasm and not a distinct part. Still Henfrey thinks it better to retain the name to express the formative stratum of living protoplasm. "In animal cells," according to Pringsheim, "partly from their small size and partly from their greater wealth of protoplasm, it is rarely possible to make a sharp demarcation between a cortical layer of protoplasm and a cell-fluid ; nevertheless there exists a difference in the constitution of the former, such that a cutaneous layer destitute of, or scantily furnished with, granules encloses the remaining more granular material. The white blood cell may serve as an example. This however is very different from a proper membrane" (Qu. Mic. J., p. 252.—1863).

With respect to the nucleus, we have seen that according to Naegeli, it is wanting in several fungi and lichens, and to Alex. Braun in the Hydrodictyon, Vaucheria, Caulerpa, &c. Also in various forms of amœbæ.

In 1853, Professor Huxley wrote a review * of the then existing literature of the cell theory, which is not only valuable historically, but important as an original contribution to it. The following are the most important points.

That the primordial utricle is the essential part of the endoplast [*i.e.*, all that is contained within the cell wall], while the protoplasm and nucleus are simply its subordinate, and, it might almost be said, accidental modifications.

That the process in cell-division, and the histological structure of plants and animals is essentially identical, and thus, for example, the chondrin wall of cartilage is the homologue of the

* Med. Chir. Review.

cellulose wall of the plant, both representing the periplastic element ; while the nuclei of the former represents the primordial utricle, contents, and nucleus of the latter. With this all who have considered the foregoing evidence will no doubt agree ; but then he adds : " In the plant the primordial utricles divide, separate, and the cellulose substance grows in between the two. In young cartilage the same thing occurs : the corpuscles divide, separate, and the chondrin substance eventually forms a wall of separation between the two. There is neither endogenous development nor new formation in either case. The endoplasts grow and divide, the periplast grows so as to surround the endoplasts completely, and, except so far as its tendency is to fill up the space left by their separation, there is no evidence that its growth is in any way affected by them, still less that it is, as is often assumed, deposited by them . . . Finally, for the notion of the anatomical independence of the cells, we must substitute that of the unity and continuity of the periplastic substance in each case."

Professor Huxley directs his attention primarily to the development side of the question, putting in the foreground the gradually increasing differentiation of the amorphous germ, owing to its distinctively vital properties. So far, again, all will now agree, but he goes on to say that the differentiation into endoplast and periplast (i.e., a nucleated cell) is the first step, and by the growth and differentiation of both these, tissues are formed. " There is no evidence whatever that the molecular forces of living matter (vis essentialis of Wolff, or the vital forces of the moderns) are by this act of differentiation localized in the endoplast to the exclusion of the periplast, or vice versâ. . . . So far from being the centre of activity of the actions, it would appear much rather to be the less important histological element. The periplast, on the other hand, which has hitherto passed under the name of cell wall, contents (?), and intercellular substance, is the subject of all the most important metamorphic processes, whether morphological or chemical, in the animal and in the plant. By its differentiation every variety of tissue is produced ; and this differentiation is the result, not of any metabolic action of the endoplast, which has frequently disappeared before the metamorphosis begins, but of intimate molecular changes in its substance."

In this section we perceive that, side by side with the development and modification of the cellular theory, has been growing up the recognition of the fact that in the sarcode of animals, and the protoplasm of plants, we see the simplest and probably ultimate form of living matter. Those who are guided by the light of Fletcher's hypothesis are no doubt prepared to see in this the realization of his one and only irritable or living matter. But that any such impression was general, or, indeed, had a place at all among physiologists, even in 1853, the above testimony of Huxley is sufficiently significant in the negative.

In the foregoing résumé I have purposely confined myself to works written before 1860, and given chiefly an analysis of a few of the chief memoirs on the subject, with little comment, in order that the meaning may emerge unbiassed by observations made in the light of subsequent knowledge, which is seldom the case with compendious treatises. From this it appears that the threefold form of the cell was soon given up, and that, although the dual form of the nucleated cell, without cell wall, nominally holds its place still, life has been shown to exist in the simple form of a structureless, viscid, semifluid matter common to both animals and plants. The bearing of this on the localization of the living matter in the complete individuals of the higher orders appears as yet not to have been thought of, and almost the whole reference of these facts has been made to development alone. And, in fact, Schwann's and Schleiden's chief aim was to demonstrate the existence of a "common principle of development for all the elementary parts of the organism." This has been most fully admitted by Huxley in the foregoing extracts, and he plainly states that the truly vital functions of growth and differentiation reside in all parts of the typical cell and their derivations. And this is indeed still held by the majority, although Huxley's mode of statement in the above review will hardly be accepted now, and his own opinions have undergone considerable change.

CHAPTER IV.

BEALE'S PROTOPLASMIC THEORY.

AMIDST the bewildering variety and confusion of the different opinions up to this time, we are struck with the appositeness of a remark by Leydig on the function of the instrument which has revealed so much. "We microscopists," says he, "it appears to me, find ourselves, alas, in the position of one who has been for a long time studying 'life' as he would a meadow or a wood from a distance, and fancies that if he could only get nearer, so as to see under his eye the individual plants which made up the verdant surface, he would at once attain to a better understanding of the process of growth of the plant and the fading of the leaf. Truly, he would learn many new things of interest, but the main point would remain as much a riddle as before: the same questions would remain to be answered, but they would apply now to the individual plant instead of the verdant landscape as a whole."

During the preceding quarter of a century we have been brought nearer, and seen much that is interesting; and although, even if we could localize life in one substance alone,* that would not explain its ultimate nature, yet it would be a great step in science. That, however, had not, as yet, been done by practical histologists, and the foregoing observations, which seem to us so plainly to point to one conclusion, still remained isolated, or, at most, applied to development. It seemed as if, at this stage, some master-mind was waited for, who should unravel the complexity, and reconcile the seeming contradictions of the previous results of histological research, and show practically how the whole circle of vital actions could be contained in the changes of a single structureless anatomical element, thus realizing the hypothesis of Fletcher.

Such appeared in 1860, we are glad to say, in the person of our countryman, Dr. Lionel Beale, of London. He had for years devoted himself with unwearied zeal to microscopical research on the animal tissues, using the highest magnifying powers as soon as available, and had attained to an almost unrivalled skill in, and had discovered various new methods of, the preparing objects which enabled him to analyze the structures of the textures to a point hitherto not reached by anatomists. As the result of these original researches, and no doubt taking into account the general progress of biological science in this direction as given above, he in 1859 drew attention to the significance of

* "The Protoplasm or Sarcode theory, *i.e.*, . . . that this albuminous material is the original active substratum of all vital phenomena, may perhaps be considered one of the greatest achievements of modern biology, and one of the richest in results."—Häckel, Qu. Mic. J., 1869, p. 223.

the living, moving matter which constitutes the body
of the amœba, the white blood corpuscle, the pus corpuscle, and so-called naked nuclei, which were to be
detected in animal and vegetable tissues; and described
how this matter could be distinguished from the proper
tissue.

Then in 1860 he wrote those "Lectures on the
Structure of the Simple Tissues of the Human Body,"
which were delivered before the Royal College of
Physicians in 1861; and which, I believe, are destined
to mark an epoch in the progress of Physiological
Science.   Since then Dr. Beale has gone on completing
and expanding his system and filling up the details,
and has carried it out into pathology to an extent of
completeness and consistency marvellous for the short
time as yet given, and as being the work of one man : a
fact which in itself shows he has seized on one great
and central principle, which enables him to bring into
practical harmony a vast number of scattered observations both of his own and of others.

Nevertheless, his doctrine is substantially the same
now as in 1860, and all questions of priority must be
judged of from documents existing then, and not from
the opinions of men now; for, as we shall see, a great
many of the current teachings on the subject are substantially those first distinctly enunciated by Beale,
although attributed to other sources.   It is also true
that many of the separate facts, and even theories, on
subordinate points have been anticipated by others;
but he brought them into harmony, and showed their
real meaning; and if the grand theory of the one true
living matter was, as we have seen, hypothetically

advanced by Fletcher,* yet the merit of the discovery
of the actual anatomical representative of it belongs
to Beale in accordance with the usual and right award
of the title of discoverer to him alone who demonstrates
truths by proof and fact.   The rising generation of
medical men will also recognize in Dr. Beale the title
of discoverer because he has first and consistently
pressed the doctrine forward with the perseverance of
firm conviction in text-books and medical works. The
cardinal point in the theory of Dr. Beale is not the
destruction of the completeness of the cell of Schwann
as the elementary unit, for that was already accom-
plished by others.   Nor that some constituent of the
cell—nucleus or protoplasm, or some matter analogous
—is embryologically the precursor of all tissues and
parts, for that is almost a truism considering what is
obvious in the origin of each individual in the ovum.
But that, from the earliest visible speck of germ, up to
the last moment of life, in every living thing, plant,
animal, and protist, the attribute of life is restricted to
one anatomical element alone, and this homogeneous
and structureless; while all the rest of the infinite
variety of structure and composition, solid and fluid,
which make up living beings, is merely passive and
lifeless formed material.   This distinction into only
two radically different kinds of matter, viz., the living
or germinal matter and the formed material, gives the
clue whereby he clears up the confusion into which
the cell doctrine had fallen, and gives the point of departure for the theory of innate independent life of

* Dr. Beale has informed me that he had not seen Fletcher's
" Rudiments of Physiology" before 1869.

each part, which the cell theory had aimed at but failed to make good. The one true and only living matter, called by Beale germinal matter, or bioplasm, is described as " always transparent and colourless, and, as far as can be ascertained by examination with the highest powers, perfectly structureless; and it exhibits those same characters at every period of its existence: * * * it would not be possible to distinguish the growing, moving matter which was to evolve the oak from that which was the germ of a vertebrate animal. Nor can any difference be discerned between the germinal matter of the lowest, simplest epithelial scale of man's organism, and that from which the nerve-cells of his brain are to be evolved."

The living matter of Beale corresponds to the following histological elements of other authors: The viscid nitrogenous substance within the primordial utricle, called by Von Mohl, Protoplasm; the primordial utricle itself, in Naegeli's sense of that term, viz., the layer of protoplasm next the cell wall: the transparent, semi-fluid matter occupying the spaces and intervals between the threads and walls of those spaces formed by the so-called vacuolation of protoplasmic masses: the greater part of the sarcode of the monera, rhizopoda, and other low organisms; the white blood corpuscles, pus corpuscles, and other naked wandering masses of living matter; the so-called nucleus of the secreting cells, and of the tissues of the higher animals, and many plant cells; the nuclei of the cells of the gray matter of the brain, spinal marrow, and ganglions, and the nuclei of nerve fibres.

The term of true living or germinal matter can

never be given to the following parts, although to some of them the word protoplasm has been erroneously applied, viz., the cell-wall of plants or animals, however delicate or gelatinous; the threads or filaments and walls of the vacuoles within protoplasmic masses or cells; the wall of the primordial utricle; the true fibrous, connective, elastic, bony, or other tissues generally included among the living parts of animals; even the proper contractile fibre of the muscles, the radiating fibres of the caudate nerve-cells, and the outer coat of those cells, besides the nerve-fibres in general; the hard parts of epithelial cells, and all liquid secretions; the cilia; the tissue of cuticle, hair, nails, horn, and all analogous parts in plants; the granules in sarcode; all colouring matter; and, lastly, all pabulum, including the fluid part of blood, lymph and chyle, and corresponding matters in plants.

In short, the name of bioplasm, given by Beale, or protoplasm* (in a restricted sense, as it will probably

---

* Since the unity of the ideal living matter has been recognized, and its characters sharply defined, the question of its name assumes importance. The matter originally termed germinal matter by Beale, corresponds more or less closely with the plasma of Häckel, the protoplasma of Mohl, the sarcode of Dujardin, the cytoplasma of Kölliker, the cell-stuff, or formative matter, of various authors; but the word protoplasm as defined by Kühne is the most generally used to express the idea of living matter as nearly as is yet adopted by others in the sense of Beale. But as on various occasions that word has been used in a loose way, and applied to objects which have no title to vitality, Dr. Beale objects to it. And since, for several reasons, the original term, germinal matter, is inconvenient, he proposes a new name instead of attempting to restrict the word protoplasm to one accurately-defined meaning. He says, "The name I propose to give to the living or self-increasing matter of living beings, and to restrict to this, is bioplasm. Now that the word biology has come into common use, it seems desirable to employ the same root in designating the matter which it is the main purpose of biology to investigate. Bioplasm involves no theory as regards the nature of the origin of the matter; it

be ultimately accepted by biologists), as indicating the
ideal living matter, cannot be given to any substance

---

simply distinguishes it as living. A living white blood-corpuscle is a
mass of bioplasm, or it might be termed bioplast. A very minute
particle is a bioplast, and we may speak of living matter as bioplasmic
substance" (Qu. Mic. J., July, 1870.) If this name had been given at
first, and at the same time the idea as rigidly defined, probably it
would have been universally adopted, but whether it ultimately will
be now since other words are current already, no one can tell. With
respect to the propriety of the word bioplasm to express what is meant,
an excellent classical scholar—Mr. Scott, of Birmingham—to whom I
submitted the question, says, "There is nothing in the word itself to
indicate that the thing formed is also itself living ;" hence it would apply
equally to the formed material. Häckel ("Generelle Morphologie,"
i., 1866, p. 276) says τὸ πλασμα signifies properly the thing formed,
and the forming material would be better designated plasson, from
τὸ πλασσον. Accordingly, more recently, he calls the substance of the
cytodes "plasson," while the substance of the nucleated cells is named
protoplasma (Die Kalkschwämme, 1872). This would simply amount
to inventing another new name for the ideal living or germinal matter,
and restricting the word protoplasm to a particular variety of living
matter. The name given by Häckel to what corresponds with Beale's
bioplast, is plastid, which is a very convenient word if bioplast should
not be ultimately adopted. As observed by Dr. Sharpey, the word
protoplast having been already taken up for such a widely different
signification, it is not available in physiology. "On the whole, as
remarked in my former work, it is most probable that the term protoplasm will still be retained in this country to express the idea of
Beale's germinal matter and bioplasm, as it is so accepted by Dr.
Sharpey, whose calm and solid judgment reveals to us, as it were, the
verdict of posterity." Dr. Beale himself seems to acquiesce in this,
or he says in his last edition of "Protoplasm," 1874: "It is, I think,
possible that after some years have passed, *protoplasm* may be restricted
to living matter only. The term will then become synonymous with
bioplasm, or living matter, in which case the latter words may be
given up" (188). I will, therefore, continue to use as synonymous
with the ideal living matter the words "irritable matter," "living
matter," "diffused ganglionic nervous matter," "germinal matter,"
"bioplasm," "protoplasm;" and for adjectives, protoplasmic, bioplasmic, or living; and for an individual mass, protoplasmic mass,
bioplast, and plastid. However, as every visible mass, or plastid,
contains, in addition to the ideal living matter, some pabulum and
some formed material, besides water, I would suggest that the gelatinous matter forming the bulk of many of the lowest and so-called unicellular organisms should still, as originally, be called *sarcode*, while
the word *protoplasm* should be strictly reserved for the true ideal
living matter.

displaying rigidity in any degree, from the softest gelatinous membrane up to the hardest teeth-enamel; nor to anything exhibiting a trace of structure to the finest microscope; nor to any liquid; nor to any substance capable of true solution.

Thus "nothing that lives is alive in every part," but as long as any individual part or tissue is properly called living it is only so in virtue of particles of the above-described protoplasm freely distributed among, or interwoven with the textures so closely that there is scarcely any part 1-500th of an inch in size but contains its portion of protoplasm. Thus we see realized the hypothesis of Fletcher, that all living action is performed solely by virtue of portions of irritable or living matter interwoven with the otherwise dead textures. According to Beale, "of the matter which constitutes the bodies of man and animals in the fully-formed condition, probably more than four-fifths are in the formed and non-living state. All this was, however, living at an earlier period of existence." This is on an average, for some tissues contain much less living matter; the bones, for example, only 1-20th, and some textures, when old, not more than 1-100th.

Vital Properties of Protoplasm.—The chief vital properties of the living matter, or bioplasm, are thus described by Beale: " It alone is concerned in development, and the production of those materials which ultimately take the form of tissue, secretion, or deposit, as the case may be: and of producing matter like itself out of matters differing materially in composition, properties, and powers. Upon it all growth, multiplication, conversion, and, in short, life, depend.

By its agency every kind of living thing is made, and without it, as far as is known, no living thing ever has been made, or can be made at this time, or ever will be made."

"The difference between germinal, or *living matter*, or bioplasm, and the *pabulum* which nourishes it, on the one hand, and the *formed* material on the other, is, I believe, *absolute.* The pabulum does not shade by imperceptible gradations into the living matter, and this latter into the formed material; but the passage from one state into the other is sudden and abrupt, although there may be much living matter mixed with the little lifeless matter, or *vice versâ. The ultimate particles of matter pass from the lifeless into the living state, and from the latter into the dead state suddenly.* Matter cannot be said to *half-live* or *half-die.* It is either *dead* or *living, animate* or *inanimate;* and formed matter has ceased to live" (Protopl., 3rd edit., p. 185.)

It is unnecessary to multiply quotations on this subject, as further illustration of what is deemed specifically vital action will come in incidentally as we proceed. Indeed, the majority of vital properties are not capable of being singled out for separate demonstration—the whole of physiology being, in fact, nothing but these properties in action—but there is one which is palpable to the senses, which may be noticed, viz., that which gives the power of apparently spontaneous movement. These protoplasmic or bioplasmic movements have attracted the attention of all observers; but particular stress is laid on them by Dr. Beale as an important element in vital functions, and also as an evidence of the essentially distinct nature of vital power.

The movements of protoplasm masses when not confined by a rigid cell wall, as seen in the white blood corpuscle, mucus, or pus-corpuscle, the amœba, &c., are graphically described and frequently dwelt on by Dr. Beale; but, as these are so well known now, I will not repeat them, but only notice those movements instanced by him as serving a purpose, and showing a power, not explicable, as he thinks, by the properties of matter and force. In nutrition, he observes, " This formless living matter moves forwards, burrowing, as it were, into the nutrient pabulum, some of which it takes up as it moves on. It is not pushed from behind, but it moves forwards of its own accord. In a similar manner, the advancing fungus bores its way into the material upon which it feeds, and the root filament insinuates itself into the interstices between the particles of the soil where it finds the pabulum for its nutriment." " The tree grows upwards, against gravity, by virtue of the same living power of bioplasm. In every bud portions of this living matter tend to move away from the spot where they were produced, and stretch upwards or onwards in advance. No tissue of any living animal could be formed unless the portions of bioplasm moved away from one another." The movements in the cells of the leaves of Vallisneria, Chara, or Anacharis, and in the hairs of Tradescantia, are of the same nature as those of the mucous corpuscle and the amœba.* The movements in

* In 1855, Unger (" Anat. and Phys. der Pflanzen ") says, " The proximate cause of the movements of the sap in the cells " (of plants)

pus-corpuscles, and other morbid products, are of the same nature. It is also by similar movements that detached protoplasm masses which form the germs of contagious diseases "climb, as it were, through still, moist air, just as the amœba and certain other living particles are capable of climbing in any direction through water which is in a state of perfect rest. Minute particles, possessing their inherent powers of active movement, can insinuate themselves into the slight chinks in fully formed tissues, in every part of the body, and may easily make their way along the crevices between the protective epithelial cells into the tissues beneath, and thus through the thin walls of the smallest vessels into the blood."

This power of movement, Dr. Beale thinks, is also instrumental in producing the peculiar form of certain tissues in the higher animals, e.g., in "the formation of the elastic cartilage of the epiglottis, it seems probable that each mass of bioplasm revolves while it forms delicate fibres, which accumulate, and at length appear to be arranged concentrically round the space in which it lies," as the caterpillar spins its cocoon.

"is to be sought neither in diosmosis, nor in the action of the nuclear vesicle, nor in any mechanical contrivance such as cilia, but it lies rather in the constitution of the self-moving protoplasm, which, as an especially nitrogenous body of the nature of that simple contractile animal substance called sarcode, produces the rhythmically advancing contraction and expansion." "If we compare the sarcode substance of the lowest animals, such as the Rhizopods, with the protoplasm as it usually presents itself in plants, the correspondence of both in form, contraction, and activity is, in fact, very surprising." This is subsequently confirmed by the observations of Max Schultze on the Foraminifera and the Actinophrys Eichornii. He concludes that the visible movement of the granules has its seat in the substance of the contractile protoplasma itself, and not in fluid contained in, and set in motion by, the contraction of the protoplasm as supposed by Brücke and Heidenhain ("Protopl. der Rhiz," 1853 p. 66). This is now universally admitted.

The formation of the spiral fibres of the ganglion cells of the sympathetic nerves of the frog is explained in the same manner.

The apparent spontaneity, and the power of portions of this transparent almost fluid substance, with so little cohesion, and no structure and mechanism, moving in advance of other portions, against gravity, he thinks it impossible to account for without the addition of a power beyond that of matter and force.

But, in opposition to most physiologists, he denies that these movements can be the cause of muscular contraction, for the following reasons :—muscular contraction is a mere alternation of movement, limited in direction as well as regards the degree of change. In this act the particles move in a direct line, and it is not possible for any particle to get before another particle. On the other hand, the protoplasm may move in any direction, and its movements are so varied that the same mass probably never twice in its life assumes the same form; also, one portion can move in advance of another portion, and an entire mass may move onwards for a distance equal to its own diameter. Other reasons will be given in chapter viii., showing that the muscular fibre does not contain protoplasm.

Ciliary motion he holds to depend on protoplasmic action, but only secondarily, owing to currents produced thereby.

The protoplasmic movement, independently of nerve influence, is generally admitted, and also the locomotive powers of individual plastids, and so-called wandering cells, which may creep through holes and change their shape (Recklinghausen, Engelmann), and pass through the walls of the vessels.

(Waller, Beale, Cohnheim), and take up foreign bodies, and account for various phenomena. All this is distinctly recognized; but the majority are also inclined to attribute muscular contraction to the protoplasmic movements, and distinctly assert that protoplasm is contained in the muscular fibre. Hermann[*] says, "We may therefore lay down the proposition that movements (in the sense of mechanical work), in all cases, only manifest themselves where protoplasm occurs" (207).

All protoplasm, besides the various partial movements above described, is capable of contracting into a ball under electricity and various stimuli, and it is taken for granted that in mus-cular contraction something similar is the efficient cause of the movement, especially as the same stimuli cause contraction of the protoplasm and the muscles. On carefully reviewing all these statements, I find them very weak as arguments that muscular contraction depends on protoplasmic movements, and it rather appears as if the question had never yet been fairly raised than that it had been settled. On the other hand, on the question of spontaneity, other authors are unanimous in stating simply that the stimuli are unknown, and on this point, I think, Dr. Beale has made out no case; nor, in his assertion of some hyper-physical cause for the movements, is he more fortunate, for he admits the impossibility of even living matter creating force; while, according to his own showing, there is always a sufficient change of matter going on to account for evolution of the necessary force. Nor have we any proof that the slight viscidity of the protoplasm does not show sufficient cohesive force to account for the movements of bioplasts, which never surpass that sufficient to sustain their own weight,[†] and a drop of water has enough for that. This slight cohesive force, however, is a most cogent argument against the protoplasmic theory of muscle-work.

[*] " Grundriss der Physiologie."

[†] In opposition to this, it is urged by Dr. Beale that a growing mushroom will raise a stone hundreds of times its own weight; and that a growing root will split a tree, &c. ; but in all these instances molecular forces are in play, which depend on the chemical changes involved in deposition of structure and in physical forces exerted by that structure afterwards. A wedge of dead wood inserted in a rock dry, will split it when moistened by the mere force of capillary attraction.

On the Chemical State of Protoplasm.—On this point, the similarity, or, indeed, identity, of the views of Beale with those of Fletcher (see p. 6), is surprising. A few quotations will show this :—

" Of the chemical composition, and of the actual state, speaking in a physical sense, of the living matter, we as yet know nothing. Nor have we even been able to hit upon any method of investigation which offers a fair chance of enabling us to ascertain the knowledge we so much desire to gain. If we attempt to analyze living matter it becomes changed. We examine not the actual living, growing matter itself, but the substances which result from its death. The facts of the case do not permit us to conclude that the materials we discover actually existed during life. On the contrary, the evidence is conclusive that the substances we test, and examine, and handle, did not exist in the condition or state known to us until the matter of which they consisted had ceased to live" (" Protoplasm," third edition, p. 32). Again, " It seems probable that during this temporary living state the elements do not exist in a state of ordinary chemical combination at all. These ordinary attractions, or affinities, seem to be suspended for the time." And again, " To assert that living matter is ' protein,' or ' albumen,' is to assert that which never has been, and never can be, proved, and all arguments based upon such assertions must be discarded."

The essential complexity of the state of aggregation of the living matter is here recognized.

" There can be no doubt that the smallest particle of living matter is complex. It is impossible to conceive the existence of a *living particle*, consisting of a simple substance only, as *iron, oxygen, nitrogen*, etc.; for living involves changes in which different elements take part. It appears to me that the term *living atom* cannot with propriety be employed, because *living matter* is of complex composition, while the idea of an *atom* seems to involve simplicity of constitution, if not indivisibility" (" Protoplasm," third edition, p. 281).

It is also recognized, and frequently insisted upon, that all the chemical elements which are contained in any tissue must have previously passed through the living state; hence the protoplasm of the different parts must contain them all, at least for a time. The analysis of all the tissues and products of vital action, animal and vegetable, gives us the sum of all the chemical elements which must enter into the composition of protoplasm as a whole. But the particular form of compound provided by each kind of protoplasm in the slow action of nutrition and secretion is not indicated at all in the composition of the living matter changed rapidly into ordinary chemical compounds, and very little difference can be detected as yet by chemical analysis between the products formed by the sudden death of the different kinds of living matter. These products are thus described by Dr. Beale :—

"When the life of a mass of bioplasm of any kind is suddenly cut short, lifeless substances, having very similar properties, result. These substances belong to four different classes of bodies. One separates spontaneously soon after death; another is a transparent fluid, which is coagulated by heat and nitric acid; the third consists of fatty matter; and the fourth comprises certain saline substances. When a mass of bioplasm dies, it is resolved into :—1. fibrin; 2. albumen; 3. fatty matter; and, 4. salts. These things do not exist in the matter when it is bioplasm, but, as the latter dies it splits up into these four classes of compounds" ("Bioplasm," p. 11).

The cardinal points here shown are that no binary or ternary compounds alone can form living matter, but all protoplasm must contain the four so-called organic elements, with sulphur, phosphorus, and some other chemical elements, which may vary, and that the most notable phenomenon which attracts attention at death is the formation of a spontaneously coagulable substance. The coagulation of the fibrin of the blood, and the stiffening of the muscles, from formation in those of the same or of an allied substance, have long been objects of common observation. Kühne holds the coagulable substance of the muscles to be somewhat different from fibrin, and names it myosin, but he maintains that the coagulable matter of the universally diffused protoplasm depends on myosin, and that it is on this as well as the stiffening of the muscles that the *rigor mortis* depends.

That the final stage of the formation of fibrin is a mere chemical combination is now established; but the precursors of it are produced by the death of living matter, although, probably, not both at the same time. Dr. Beale does not pronounce a definite opinion on the nature of the coagulable matter of protoplasm.

An acid reaction is shown by the sudden death of all kinds of protoplasm. The totally different conditions under which a compound, capable of being also made in the laboratory, is made by the living matter, is one of the strongest arguments for the radically different chemical state of the living matter from that in which it subsists in the inorganic condition. This would require a volume to illustrate fully, so it may be merely alluded to as an argument.

On the Physical State of the Protoplasm.—" The living matter is in all cases perfectly clear and transparent. It never exhibits structure; is invariably colourless." The supposed granules are merely fatty matters developed at death, or foreign bodies if seen in living amœbæ, or such living corpuscles. From its transparency and total absence of structure, it has been mistaken for mere passive fluid, occupying a space or vacuole. It is a semi-fluid, slightly viscid substance, like gum or syrup, but not susceptible of true solution. "The elementary particles of germinal matter are invariably spherical, although the masses compounded of them very much in form. There is reason to believe that the spherical particles are themselves composed of spherical particles, and so on to a minuteness far beyond that which it is possible to realize." If we could see into such a mass, says Dr. Beale ("Prot.," p. 277), "in each little spherule the matter would be in active movement, and new minute spherules would be springing into being at its central part. Those spherules already formed would be making their way outwards, so as to give place to new ones, which continually arise in the centre of every one of the animated particles." Each spherical particle is free to move in fluid, and the spaces which, we must conclude, exist between the spherical particles of living matter are probably occupied by fluid." This fluid contains, in solution, pabulum, or matter about to become living—substances which exert a chemical action,* but do not form a constituent part of the living mass—and "substances resulting from the changes ensuing in particles which have arrived at the end of their period of existence, and the compounds formed

* This probably includes those soluble matters suitable for maintaining the fluid in a proper osmotic condition, and also which may act as stimuli, which last Dr. Beale seldom alludes to.

by the action of oxidation of these" (p. 281). From the
above physical characters, common to all protoplasm,
it *appears* to be precisely the same in all living
structures.   "The germinal matter of an embryo
resembles that of the tissues of the adult, and the
germinal matter from the most inveterate morbid
growth could not be distinguished from that of a
healthy tissue ;" nor can that of the lowest fungus
and of the brain of man be distinguished, by microscopic or other physical examination.

*The Relation between Beale's Protoplasmic Theory
and the Cell Theory.*—From what has been said, it is
plain that the idea of the cell, either in its triple or
double form, as the ultimate morphological or physiological unit, is simply abolished altogether, already in
the year 1860.   Nevertheless, as the cell, in some form
or other, is an anatomical fact, it is necessary to say a
few words in elucidation of the correspondence between the current nomenclature of parts already
known and that of Dr. Beale.   The latter would still
retain the word cell as a convenient term to indicate
the anatomical unit, or elementary part, with the
proviso that there must always exist the two radically distinct portions—viz., the bioplasm and the
formed material, however complicated and varied the
last may be.   In some so-called cells there is nothing
but bioplasts and intercellular substance, which latter
corresponds to the cell wall.   In other cells there is no
solid cell wall, but only the bioplasm, and the formed
material is represented by the secretion—e. g., the
liver cells.   In others—e. g., the so-called secreting
cells—there is, besides the bioplasm, the secretion and

a cell wall, within which the latter accumulates. This applies to the flask-like cells, the fat cells, and the starch cells, &c., and also to the more complicated ciliated cells. There is no difficulty of seeing how this applies to continuous tissues, and all matters enumerated as dead at p. 45; and there is no necessity of straining the theory to make it fit, as was the case with the old cell theory, where nothing like the transformation of a typical cell could be seen in most cases. But the question of the nucleus is more difficult; and here the definite conception of Beale brings order and clearness out of the confusion and obscurity which prevailed. In the first place a number of objects are simply falsely called nuclei, and are nothing but oil-globules, &c., resulting from post-mortem changes in germinal matter; or are granular and other foreign bodies. Next, in many independent organisms of low order, such as algæ, monads, and infusoria, the so-called nucleus is really an organ, though its use is not yet determined.* Again, the term nucleus is really

---

* That the nucleus is not an essential element, but a special organ, performing, when present or visible, some more or less manifest part in the process of reproduction, becomes more and more apparent as the lower forms of life are more efficiently studied. It is now well known that in *Paramœcia* the reproduction is partly by fission and partly sexual. The sexual mode, when carefully studied with sufficient power, leaves no doubt as to the fact that both the nucleus and the nucleolus are special organs—the nucleus being the ovary, and the nucleolus the testis; for it is *only* after the mingling of the elements contained respectively in these that any sexual multiplication takes place. Van Beneden has studied the development of Gregarinæ from pseudo-filaria, which are at first simple threads of protoplasm. After a short period of quiescence a nucleolus forms—but by what means this is distinguished from the nucleus is not stated; but it takes its rise in elements previously contained in the protoplasm. Then the nucleus is formed; but its function is said to be unknown. But the Rev. W. H. Dallinger, who has given careful study to the Gregarinæ from the earth-

the bioplast, or representative of the only living matter, or cytoblast, of the part. This is the case with the continuous tissues, the epithelial cells of both kinds, the fat and starch cells, the capillary walls, the bone cells; in new growths, and in young and quickly-growing parts, &c. There remains, then, only the so-called nuclei and nucleoli, which exist within the protoplasm, to which the term ought properly to be applied. Dr. Beale dissents from the theory that these bodies are the parts first formed, and which cause the rest to be deposited round them; and also that they are the sole agents concerned in reproduction. On the contrary the protoplasm is formed first, and these appear in it afterwards. Often subdivision of the protoplasm takes place, and nuclei appear in a portion after it has been detached. They have the same composition as the protoplasm, and are merely new centres of more intense vital activity. "Bioplasm in a comparatively quiescent state is not unfrequently

---

worm, informs me that the cyst from which the pseudo-naviculæ emerge remains finely granular, and never breaks up into pseudo-naviculæ until the nucleus has opened, and its contents are distributed throughout the sac. The same observer has shown that in the amœba the nuclei differ sexually, which has been recently confirmed by Greef; and that the transfusion of their contents is needful for multiplication by this method ("Proc. Lit. and Phil. Soc., Liverpool," 1871—2, pp. 297, 8). In the vorticella, besides the methods of increase by fission and gemmation, there is also a true sexual mode of multiplication, in which the vorticella encysts, the cilia and pedicle disappearing, and the nucleus, which with very high magnification is shown to have an enclosed nucleolus, rupturing soon leads to the breaking up of the forms into a number of sub-spheroidal germs, which are set free by the breaking up of the cyst; and after some little time of locomotive life, these develop into fresh vorticellæ.

Many other examples might be given, but these will suffice to show that in independent organisms, the nucleus and nucleolus may be, in fact, organs whose use is known, or as yet unknown, and not essential components of living matter in the abstract.

destitute of nuclei, but these bodies sometimes make their appearance if the mass be more freely supplied with nutrient matter." "The nucleus is a new centre of growth, and within it new centres may arise. The nucleus has the power of resisting the action of conditions which would destroy the remainder of the germinal matter; so that the nucleus may retain its vitality under certain circumstances which would certainly cause the destruction of the elementary part, and this nucleus may at a future time grow, and produce an infinite number of elementary parts."* "The nuclei and nucleoli, although they are bioplasm, do not undergo conversion into formed material. Under certain conditions the nucleus may increase and exhibit all the phenomena of ordinary bioplasm; new nuclei may be developed within it, new nucleoli within them; so that ordinary bioplasm may become formed material, while its 'nucleus' grows larger, and becomes ordinary bioplasm" ("Biopl.," p. 55). "The formation of new centres within centres goes on for many series, each acquiring new powers, till at last one is evolved which forms the tissue or organ. These are all direct descendants of each other, and each retains, by inheritance, some of the powers possessed by those which preceded it." Thus the nature and functions of the nuclei are explained, and, to a certain extent, in harmony with previous views of a certain pre-eminence in this body (see p. 20).

Such are the cardinal properties of the one anatomical substance which forms the basis of life. We

---

* Beale: "Med. Chir. Review," vol. xxx. p. 210.

may now consider the chief facts and observations on which this theory is founded. The carmine-staining process may be first noticed, as it has, in addition to its own importance, the historical interest of having led the mind of Dr. Beale to his great discovery.

The discovery of the carmine-staining, which has borne such important fruits, appears to be due to Lord S. G. Osborne, and it was made known by him in a memoir read before the Microscopical Society, in June, 1856, and published in the fifth volume of the " Journal of Microscopical Science." In this he says, " It was the desire to trace out this feature of cell growth [formation of spongiole cells from protoplasm], which led me to seek the means of giving colour to the formative matter. To my delight, I now found that, whilst the mass would take up or involve the pigment, the actual cell walls and epidermic plasm would not" (p. 112). He states that the "formative matter," under which term he includes "granular and molecular matter," embodied in a thick, transparent fluid, may be coloured, and that it is this which effects the various combinations of cell contents; but he is not sure that the staining is uniform in this, or proceeds from some very minute, highly-coloured granules. He observes also that the nuclei were more highly coloured, and says, "Under every feature of cell growth, the nuclei and the aggregations of formative matter which tend to nucleolar growth, will always be found to present a much deeper colour than the formative matter in the same cell with them" (p. 114). His plate exhibits the staining of the protoplasm and nucleus of the cells of the spongioles of the wheat-root, and is very like what we are now familiar with. In one place, also, there is exhibited the closely set stained masses of the newly-formed portion ; while higher up in the same rootlet the cells are oblong, and nearly colourless, with a single deeply-coloured nucleus. The mode of staining was by letting the wheat-plant grow in spring water coloured with carmine. We see, thus, that this author makes a considerable step in advance in the protoplasmic theory, but he did not reach it ; for he speaks of the "plasm" of the cell wall secreting formative matter, and he does not attempt to found a general theory of living and dead matters on these observations.

I have repeated the above experiments by growing wheat in water to which carmine, dissolved in a weak solution of liquor ammoniæ, was added in various proportions ; and also grown in earth, and watered with the above solutions. The general results were quite confirmatory of those discovered and figured by Lord S. G. Osborne ; but, in many instances, the general tinting, especially of the spongioles, was too uniform and intense. The colouring matter had been deposited irregularly, and on both the formed material and the amorphous proto-plasm. But in some of the youngest cells, at the apex, which I have now mounted in glycerine, the distinction can be very well seen. The cell wall is seen as a transparent glassy mem-brane, while the contents are stained irregularly in different degrees of intensity, just like the yeast cells figured by Dr. Beale. In the older cells an intensely stained spherical nucleus is seen, as figured by Lord S. G. Osborne. On the whole, the distinction into living matter and formed material is not so palpable as to force itself on the attention ; and, indeed, we cannot expect that it should be as perfect as in the staining of recently dead organisms, for the currents and frequent chemical changes following vital action must disturb the regular deposi-tion of the alkaline colouring matter.

The next step is thus narrated in Dr. Beale's own words :—

" Not very long after the appearance of Lord S. G. Osborne's paper in the 'Quarterly Journal of Micros. Science,' on the wheat plant, 1856, I began staining animal tissues, as Gerlach and others had done before me. At this time I was injecting tissues with glycerine, and preserving them in the same fluid. Naturally, the carmine tinting was tried with glycerine. Tissues so prepared were very favourable for dissection ; and, as many were prepared in the same manner, they were examined in series. The distinction between germinal matter and formed material became at once evident, and all the difficulties and confusion respecting cell wall, contents, intercellular substance, the division of tissues into cellular and non-cellular, were removed. Then followed the comparison of the several tissues at different ages, and no other conclusion than that which I formed was left for the mind to accept. My pupils received these views at once, and no one who saw series of specimens could refuse to give up the old ideas" (MS. letter).

The rationale of this process is as follows. We have seen that, with Beale, as with Fletcher, the protoplasm or living matter is in a state of combination totally distinct from that which it assumes at the moment of death. When that is rapid, as we have seen, there result formless masses of the proximate principles in which only the characters common to the proximate organic products of all living matter are discernible, while the minute differences of it in the living state are not recognizable by chemistry at present. One character common to all kinds is "an acid reaction."

Dr. Beale says (" Microscope," 4th edit., p. 107), " The living matter possesses an acid reaction, or, to speak more correctly, an acid reaction is always developed immediately after its death." But as in life all bioplasts contain some products of vital action, which goes on incessantly, we may expect that in parts in which rapid activity goes on the living matter may be described as acid. Accordingly, it is stated by Ranke (p. 11, " Physiologie," 1872) :

" In the animal cell also the chief activity of cell chemistry seems to originate in the nucleus. We see the vital activities of the organs run their course with the formation of organic acids, e.g., lactic acid, the production of which is copious in proportion to the heightened activity of the organs. Hence we

see the neutral or slightly alkaline reaction of the muscular or
nerve tissues give place, under strongly exerted activity, to an
acid one.   These chemical transformations of the cell contents
originate, as it appears, for the most part, in the nuclei, which,
in the living cell, exhibit constantly an acid reaction " (Beale,
Kölliker, J. Ranke), "in contradistinction to their alkaline en-
vironment.   This acid reaction is made known by the property
of the nucleus to colour itself red quickly and permanently in
neutral carminate of ammonia solution by fixation of the car-
minic acid."*

If, therefore, the operation of staining is performed
sufficiently soon after death, we have a most valuable
test of the parts of the organism that were living at
the moment of somatic death.   If, however, a certain
time is allowed to elapse, decomposition sets in, and
this distinction is lost.   " Certain other precautions in
practice are necessary, and the density and composition
of the colouring fluid must be slightly varied in special
cases.   But it is necessary that I should state distinctly
that if the process be properly conducted, *every kind
of living or germinal matter, or bioplasm, receives
and fixes the colour, while no kind of formed material
known is stained under the same circumstances*"
(" Bioplasm," p. 44).   The importance of such a method
in proof of the protoplasmic theory is so obvious that
opponents have not failed to question these statements,
and the possibility of tinting all sorts of substances

---

* Dr. Beale also says: "The carmine does not actually stain living
matter; and the acid, I conclude, is set free at the moment of death.
Soon after death, when decomposition begins, ammonia is formed, and
the acid neutralized, and then *my* staining will not occur; but anything may be stained, in another sense, by proper soaking. If bioplasm be carefully preserved soon after death, it may be stained, though
not so nicely, at any distant period if kept in a medium which will not
neutralize or quickly dissolve out the acid of the bioplasm " (MS.
letter).

with this same carmine solution has been said to de-
prive the test of all value. To this Dr. Beale replies
that mere steeping of tissues and other formed matters
in the solution till they are soaked through with the
colouring matter, will certainly tint all kinds of formed
matters; but the staining he means is essentially different from that mere tinting, and by observing proper
precautions it can easily be secured.

Dr. Ransome, in his paper on the " Ovum of Osseous Fishes"
(" Phil. Trans.," 1867), reports unfavourably of the process as a
means of distinction of the living and dead, for he says the
yelk sac and some other kinds of " formed material " took the
colour more quickly than the protoplasm.

And again, in his paper on the " Ovarian Ovum of the *Gas-
terosteus Leiurus*" ("Quart. Mic. J.," 1867, p. 1), he repeats his
objection, and states that the ammonia in Beale's carmine fluid
"rapidly dissolves the germinal vesicle and its contents," and
that "the granular formative yelk takes the dye with greater
difficulty than the yelk sac does, except in very young ova, and
the inner sac—a true germinal matter—does not take any
stain ; so that I cannot accept the staining of certain parts of
a structure as satisfactory evidence of the distribution of germinal matter in the tissue." In reply to this, Dr. Beale states,
at p. 85 of same volume, that the ammonia does not dissolve
the germinal vesicle or its contents, although it may precipitate
some particles from the contents, and somewhat alter its appearance ; and that in the staining Dr. Ransome has probably
failed from not observing certain precautions and modifications
necessary in practice for different objects. In proof of this, he
gives the plate of the ova of the stickleback perfectly stained
according to his method. This plate is also to be seen at
p. 306 of "The Microscope," 4th edition. I would also add,
that in the above memoir in the " Phil. Trans.," Dr. Ransome,
after saying that the yelk sac took the carmine freely, goes on
thus : " The dyeing is independent of any acid reaction of the
substance dyed, as macerated yelk sacs which had been alkaline

5

from decomposition took the dye freely." This is quite contrary to Beale's directions, and the said dyeing must have been mere tinting, or else was the proper staining of a layer of bacteria.

Although the carmine test plays such an important part, yet it must not be imagined that the protoplasmic theory rests upon it entirely, any more than the Newtonian theory rests upon the falling of the mythical apple. Among the other arguments in support of it may be enumerated the following:

The constant presence of bioplasm in all living parts of the same animal, and its absence in the dead, such as cuticle, hair, horn, &c.

The same externally-similar, structureless protoplasm occurs in all living things, plant, animal, or protist.

Coloured secretions, and particles of starch and other products, are seen to be formed in protoplasm.

"Bioplasm always exists before the formed material is produced, and the latter is never found without the former having been present" ("Biopl.," 123).

Even in the vitreous humor, apparently a completely non-living part, numerous bioplasts exist in embryonic life, and a few are found in the adult.

The proportion of bioplasm in any tissue decreases with age. In the young, growing parts it greatly predominates; then the proportion is gradually reversed as growth ceases and age advances.

The continuity of the bioplasm with the formed material in process of development.

In the formation of cartilage each bioplast divides so as to produce clusters of four or more, and these

again divide to produce secondary clusters.   The proof
that the matrix is produced by these bioplasts is given
by the fact that the quantity of matrix formed is
greater between the primary than between the secondary clusters.   If the matrix grew independently this
would not be the case.

"The activity of change of an organ or texture may
be judged of by the number of bioplasts present in it"
("Biopl.," 295.)

No process properly vital or characteristic of living
beings occurs without the bioplasm taking part in it.

The peculiar form of the fibres of the elastic cartilage
of the epiglottis can hardly be accounted for except on
the supposition that each mass of protoplasm revolves
while it forms delicate fibres.   A similar process must
be supposed in the formation of the spiral fibres of the
sympathetic ganglion cells ("Biopl.," 94).

As a summary, Dr. Beale states that his theory
would need to be greatly modified, if not entirely
abandoned, "if it could be shown that the intercellular
substance of cartilage is deposited from the blood independently of the cells; or that 'intercellular substance,' or 'cell walls,' are ever formed independently
of germinal matter; or that the matter of which the
'cell wall' consists is deposited layer *upon* layer, instead of layer *within* layer; or that the germinal matter is not, at any period of development, in bodily
continuity with the formed material; or that the germinal matter is capable of exerting an influence upon
matter situated at a distance from it, or that pabulum
does not become germinal matter, but it is merely
changed or converted into new matter by some meta-

bolic action exerted by the germinal matter, without
coming into actual contact with it or becoming a part
of it; or that cell wall or intercellullar substance possesses the power of selecting certain substances from
the nutritive fluid and converting these into matter
like itself; nay, if it can grow in and form septa, as is
described by almost all observers to take place in
cartilage" (" Qu. Mic. J.," 1863, p. 97)

Such is the basis of Beale's protoplasmic theory,
which now takes the place of the cell theory. It follows Fletcher and the non-animist school of vitalists,
and also the cell theory of Schleiden and Schwann, in
abandoning the idea of the dependence of life upon a
central overruling spirit or principle animating each
individual; and imputing it as a peculiar power or
property to quasi-independent vital units possessing
specific inherent differences in each organ and part.

The speciality of Dr. Beale's theory may be summed
up in a new addition to those aphorisms which from
time to time have marked successive epochs in the
progress of biology. We have first the dictum of
Harvey, *Omne vivum ex ovo*. Then, when this was
found not to comprehend all the modes of origin of living
beings, it was changed by Milne-Edwards into *Omne
vivum e vivo*. Again, Virchow alters that into *Omnis
cellula e cellulá*. Now, Beale adds the new and
startling aphorism, never before even hypothetically
imagined except by Fletcher, *Nihil vivum nisi protoplasma !*

CHAPTER V.

THE problem is now to account for all the vital phenomena of a complicated individual of the higher orders by the sole action of this structureless, clear, semi-fluid matter. Of course, it would be quite out of place to go into the details of the formation, nutrition, and function of the individual organs and tissues, but it may be interesting to give an outline of the general process of formation of tissues, and of nutrition, secretion, and absorption, and of the process of blood formation.

In the production of formed material from the living matter, as represented in its simplest form as cell walls, or the capsules of fungus spores, the formed material first appears as a thin film on the surface of the bioplast. Then it is increased in quantity, not by deposition from soluble matters on the outside of that film, but always by formation on what is still the surface of the bioplast, although now covered, and, therefore, all increase takes place inside of the cell wall. "The growth of an elemental unit takes place always from within." At the same time, the outer surface

may become old, hardened, and cracked.   If the cell
is destined to be cast off and perish, the bioplasm in
the inside is gradually consumed by its conversion
into formed material, and is not renewed from pabulum.   The epidermic and some other cells of the
higher animals are in this case.   But when the function of the bioplast is to be continued, its substance is
renewed from the nutrient fluid which passes through
the cell wall, or other formed material corresponding
to it.   On this plan, the nutrition of cells of epithelium
and other bioplasts of man is conducted (197).   On
this principle substantially, in fact, all growth and
nutrition are conducted.   When increase or multiplication of cells, or any other formed material, takes place,
it is always the bioplasm alone which grows, divides,
and sub-divides, and becomes surrounded by formed
material produced by it.   The formed material is
always passive, and takes no part in this process, and
never of itself grows in, or moves in, and forms par-
titions.   The process of subdivision of the protoplasm
is easy to be seen.   It is shown, among others, in pl.
vi. of Beale's "Protoplasm," third edition.   When no
new growth is going on, no further subdivisions take
place, and the consumption of protoplasm and assimilation of pabulum remain balanced.   But gradually, as
age advances, the consumption of protoplasm exceeds
the renewal, and in proportion the formed material
exceeds the bioplasm in quantity, so that, at least in
some tissues, the latter amounts only to 1-100th part.
With respect to the intimate nature of the process
of nutrition, it may be said to be co-extensive with
life itself, and the whole "investigation of it may be

narrowed to the study of the changes taking place in the transparent living matter itself, and the production of the material on its surface" ("Biopl.," p. 6). So far from the food of an animal being "a portion of the environing matter that contains some compound atoms, like some of the compound atoms constituting its tissues," as Herbert Spencer says, and which are selected and deposited, Beale, like Fletcher (see p. 10), lays stress on the fact that the pabulum in general contains none of the proximate principles of the tissue ; and, even when some such are contained in it, ready formed, every part of the pabulum is decomposed, and its elements re-arranged, and it is converted into living matter. "Every particle of matter that is to become tissue must first *pass through the living state,* and its properties, character, and composition will be determined partly by the internal forces or powers of the living matter acting upon the elements of which it is composed, and partly by the external conditions present at the time when these pass from the living to the formed state" ("Biopl.," p. 72). This does not apply only to crude food, but also to the most highly elaborated blood, which, as pabulum, is always merely dead formed material, and never can nourish in virtue of the living particles it contains.

During nutrition, and all vital action, the living matter is supposed to be in the physical state described at p. 55. The tendency of the particles to move *from* a centre, as there described, necessarily creates a tendency in the fluid around to move *towards* the centre. Thus the nutrient pabulum is as it were incessantly attracted towards the centre of each bioplast, and the newly-formed bioplasm and formed material repelled towards the circumference. Every mass, however small, of living matter contains some material in each of these three different states, and some amount of this change is incessantly going on as long as life lasts, although the rate of change may vary excessively in the active, resting, and dormant stages in which the protoplasm may exist. Thus the act of change of bioplasm into the formed material, is the act of death of the living matter, and Dr. Beale constantly uses the graphic expression that the protoplasm of this or that kind *dies into* this or that tissue or secretion.* "Every form in nature, leaves, flowers, trees, shells—every tissue, hair, skin, bone, nerve, muscle—results from the death of bioplasm" ("Biopl.," p. 10).

But not only every material product, but every action properly called *vital* depends on the same agency. "Every action in every animal, from the first instant of its existence to the last, marks the death of bioplasm, and is a consequence of it. Every work performed by man, every thought expressed by him, is a consequence of bioplasm passing from the state of life—ceasing, in fact, to be bioplasm, and becoming non-living matter with totally different properties" (*ibid.*). "The germinal matter does not secrete the formed material, but becomes resolved into it, and no

* Read by the light of Beale's theory, Fletcher's chapter on death assumes a peculiar significance. Fletcher represents death as a vital process, the last of the living actions, whereby the elements are rearranged into the chemical state; for those elements being hitherto in a totally different state, the usual chemical affinities cannot take effect on them till they are released from vital affinity by vital agency. He says of the living tissues, "their vitality cannot desert them otherwise than by a vital process" ("Life and Equivalence of Force," p. 178).

metabolic, or catalytic, or any action at a distance, can account for the process." The proofs of this are contained in general proofs of the protoplasmic theory at p. 66. And besides these, Dr. Beale shows that fluids will pass through a comparatively thick layer of formed material, and reach the germinal matter in the course of a few seconds; so there is always the opportunity given for contact with the nutriment, and conversion of some constituents of it into the living matter.

As an example of secretion by a cell, he gives the liver cell, or elemental unit. "This consists of a spherical mass of germinal matter, often containing new centres of growth (nuclei), surrounded by a considerable extent of formed material, giving to the whole an irregularly oval or somewhat angular appearance. This formed material is undergoing change upon its outer surface, and, although resulting from one kind of germinal matter, becomes gradually resolved into amyloid, fatty matter, the resinous salts of the bile, and colouring matters" ("Oxford Lectures," p. 639). "How is it possible that these various substances, not pre-existing in the blood, can be all formed and separated from it at the same moment by the agency of one and the same cell wall or nucleus acting upon it at a distance? It is surely more probable that the constituents are absorbed from the blood, and converted first into germinal matter, which then splits up into these different classes of substances" (Todd and Bowman, p. 109).

"The formation of the fatty matter occurs in this way:—In the very substance of the bioplasm, but always outside and away from the new centre or nucleus, a little oil globule makes its appearance.   It results from changes in the living matter itself.   A portion of this bioplasm dies, and, among the substances resulting from its death are fatty matter, which being insoluble remains, and soluble substances which are carried away in the blood ; starch globules and other secondary deposits, formed in the interior of elementary parts, are produced in the same manner by the death of the bioplasm.   The fatty matter does not come from the blood as fat, and deposit itself in the cell, nor is it formed by the collection and aggregation of excessively minute granules, which traverse the vascular walls suspended in serum, nor is it precipitated from the nutrient fluid after the manner of crystals ; but it invariably results from the *transformation of living matter*, and different kinds of living matter, as is well known, will produce different kinds of fat.   The properties and composition of fat in different animals differ, because the powers of the bioplasm or living matter of each animal are so different" (Todd and Bowman, p. 301.)

In a similar manner, substantially, are performed all the formative and secreting functions of life, and, as above said, all the strictly vital processes.   In the production of tissues and secretions, or formed material of any kind, whether solid, liquid, or gaseous, whether coloured or colourless, or hard or soft, soluble or insoluble, their formation depends on the relation which the elements of the living matter were made to assume towards each other during the living state. The very same elements which lived in the living matter always enter into the composition of the formed material.   The relation of the elements to each other during the living state in the different kinds of protoplasm is so definite that from the same kind of living matter, under similar conditions, the same formed substances result (" Microscope," p. 313).

Although in physical aspect all kinds of protoplasm
are indistinguishable, yet inherent, specific vital properties of different kinds, in the same individual, are
fully recognized, and repeatedly dilated on, by Dr.
Beale. These differences are displayed in the reaction
with external agencies and pabulum; and although
these may exert a certain influence on the formative
process, "it is quite certain that no conceivable alteration in external conditions will cause the bioplasm,
which was to produce muscle, to give rise to nerve,
cartilage, or elastic tissue" ("Biopl.," 127). As respects
the modifying influence of external agencies, "a temperature at which one kind of protoplasm will live and
grow actively will be fatal to many other kinds.   So,
too, as regards pabulum, substances which are appro-
priated by one form of bioplasm will act as a poison to
another. . . . The formed material resulting varies to
some extent if the conditions under which its production takes place be modified . . . but under no circumstances whatever do external conditions determine the
production of a texture higher than that which the
germinal matter was destined to produce originally."
   This is the same as Fletcher's doctrine of specific
irritability residing in each separate tissue and part in
virtue of its specific kind of irritable matter, and,
when taken in conjunction with the doctrine of specific stimuli, it has a most important bearing on the
whole of physiology, pathology, and therapeutics.
   In Dr. Beale's theory thus, with the exception of a
few chemical changes that may take place in the
formed material after its production, there is no such
thing as the passage of materials from the blood, and

their aggregation into tissue or secretion, by the mere chemical action, or by any catalytic or metabolic action of the living parts in the neighbourhood, far less by any spontaneous power in the liquid blood itself; but, in all cases, every particle must first enter into and become part of the living, semi-solid matter itself.

Such is the marvellously simple and beautiful theory of Beale, and the central point on which his whole new system of physiology turns, and which will, no doubt, make his name imperishable. It gives to our hand a plain solution of the difficulties which weighed upon the materialist* theory of life. For, action at a distance is only conceivable by means of force transmitted by some suitable medium, while no possible variety of what can be properly called force could exert the formative powers characteristic of life. This subject will be discussed in a subsequent chapter, but, in the meantime, we may notice here that it has relieved Fletcher from the dilemma in which he was placed. The dilemma is thus stated in "Life, and the Equivalence of Force," p. 163 :—

He speaks constantly of the vital functions of the several tissues, while, at the same time, life is, with him, merely a series of actions, depending on the properties of the substance in which they take place, and, therefore, which cannot be communicated or conveyed in any degree. With him, as regards the higher animals, the parenchyma or capillary tissue was "the seat of all the molecular actions of the body," thus including secretion and nutrition, not only of all other tissues, but of the ganglionic nervous matter itself. "But the ganglionic nervous tissue, assuming this as the immediate seat of the property in

* I mean in its physiological sense only, as will be explained afterwards.

question, is, of course, continually renewed and consumed, like all the other tissues of the body, by molecular processes, the seat of which is the parenchyma, which thus effects incessant changes in this tissue, not only as interwoven with the substance of every other organ, but also as entering into the composition of the minute vessels of which itself consists." How can the parenchyma, or the capillary walls, take any vital part in the formation of the substance, whose presence alone constitutes their vitality, and yet which property cannot be transferred to the smallest distance? Speaking in the present language of the force-doctrine, the vital power is not the offspring of a force, and, therefore, cannot be conveyed or transferred through other kinds of matter, or the interstellar ether, so as to operate at a distance, but it is the result of an affinity proper to one particular aggregation of matter alone, and can, therefore, only act upon particles within that; nor can any catalytic action take place between portions of matter in the vital and chemical states of combination respectively.

It is obvious how the discovery of the actual living matter and the theory of Beale remove the dilemma of Fletcher, and render intelligible and comparatively simple the view which defines life as the molecular actions or changes taking place in a substance anatomically one and chemically compounded in a manner wholly *sui generis*. But, although strongly prepossessed in favour of Beale's theory, having waited for it all these years, I cannot but allow that it is opposed by certain difficulties not yet explained satisfactorily. For instance, it is stated that every tissue or secretion, solid, liquid, or gaseous, consists of elements which were immediately before in the living state and formed part of that protoplasm which dies into the said secretion. Now, the protoplasm is always a most complex compound of from five to seven elemental constituents, and among them, invariably, nitrogen. Likewise, the passage from the living to the dead state is abrupt, and with no intermediate stage. How then can a portion of protoplasm be said to die into a binary or ternary compound, which arrange its elements as you please never could, by any possibility, have been in the living state ? Starch, sugar, and other carbohydrates, and the hydrocarbons, and even carbonate of lime, silica, carbonic acid, and water, are said to result from the death of protoplasm. With respect to some of these, Dr. Beale considers that the taking up of oxygen at the moment of passing from the vital to the chemical state, and subsequent changes of the matters originally formed by the protoplasm—which in certain cases, e.g., the secretion of the epithelial cells of the kidney, no doubt take place—may account for the reduction of formed material to simpler compounds. But this cannot account for the production of the great bulk of the ternary or binary compounds, the starch and fat, for example, which appear in the middle of the protoplasm of the cells. Dr. Beale hardly meets the difficulty in the following words, and it seems as if he were unwilling to attempt merely speculative explanations, trusting to the evidence in favour of the protoplasmic theory as a whole, and leaving it to time to furnish the explanation of the how of the process.

"We know that the nutrient matter makes its way to the very centre of the living particles, and that it there becomes changed. Certain of its elements are re-arranged, and the material particles immediately acquire powers they never possessed before. Then begins a series of orderly changes, very wonderful. During the time that the matter lives, its elements are probably arranged and re-arranged many times, the proportion of some being reduced, and that of others increased, so as to prepare for the formation of molecules of great complexity as regards arrangement, though composed of very few elements" ("Prot.," 3rd edit., p. 280).

Professor Wyville Thompson, who considers Beale's view as open to insuperable objections, thinks it more probable that—

"Protoplasm, the substance of which is endowed with peculiar vital properties, has always the same composition, and that it acts simply by catalysis, inducing, under certain known laws, decomposition and recombination in compounds which are subjected to its influence, without itself undergoing any change, absorbing the nascent products of combination and decomposition, and recombining them and reserving them with reference to the development or maintenance of the organ to which it gives its life" ("Nature," May, 1871).

To this I object that such actions imply infinitely more than catalysis, which, in fact, is a purely chemical operation requiring a definite intermediate compound to be formed by double decomposition, and this resolved again, as is the case in inorganic chemical catalytic processes, such as the manufacture of sulphuric acid and of chlorine. And if a similar process, not as yet chemically explained, takes place with the nitrogenous diastase in the conversion of starch into sugar, still we have no reason to suppose the attitude of the molecules to each other less definite. But diastase possesses no power of self-renewal or growth, or any other of the properties enumerated in the above sentence, therefore catalysis explains nothing of the very points most requiring explanation, and which must be presumed to depend on a complexity of molecular constitution far surpassing anything known in chemistry—in fact the state *sui generis* of Fletcher —and which is quite incompatible with the simple attitude of double decomposition essential for catalysis.

I may suggest that the difficulty would be clearly expressed by the statement that for every molecule of simpler formed-material a portion of protoplasm must die and split up, sufficient to have maintained that in the living state. If a particle of starch is deposited so much protoplasm must die as contained its elements in the living state; not that the starch alone died out and left the remaining elements of its protoplasm still living. Hence it must follow that a much larger quantity of protoplasm must die than the corresponding quantity of the specific formed-material, and of necessity a number of by-products must always be formed in every act of nutrition and secretion, and, in fact, in every vital act. And these must have a compensating or complementary character.

Dr. Beale seems to suggest this in the following words : "When germinal matter becomes resolved into formed material, other compounds are produced besides the special ones which characterize that particular kind of germinal matter " (Todd and Bowman, 104). And again speaking of the vegetable cell; "in all these cases the formation of the peculiar and characteristic substance which accumulates, is accompanied by the

formation of soluble and gaseous matters which escape"
(Ibid. 108).

It will be the business of organic chemistry to test
this, and thus confirm or disprove Beale's theory. At
present things are hardly far enough advanced for a
complete discussion of the question, but reasons are
not wanting for believing that such by-products are
always formed, and then reabsorbed and decomposed
and converted again into living matter, and from that
to pabulum, either on the spot or through the complicated process of blood formation possessed by the
higher animals, till finally nothing but highly oxidated
and effete matters are excreted by the emunctories.
In muscular work, for example, most probably a portion of nitrogenous protoplasm, corresponding to the
force evolved, dies and splits up in every muscular act
—for the notion that force is evolved by the direct
combustion of starch and fat is, no doubt, a gross
chemical figment—yet there is no increased elimination of urea, for of the long list of products known and
unknown in addition to lactic and carbonic acids, the
bulk of them, and probably nearly all the nitrogenous
ones, are re-converted into blood, and the amount of
urea secreted depends upon the character of the diet
and not on the amount of muscular work, as will be
more fully shown in chap. viii. In the low organisms,
whose business is destruction rather than economy,
these by-products are more plainly seen. For example,
in the alcoholic fermentation of sugar, which is a vital
process performed by the protoplasm of the yeast cells,
besides alcohol and carbonic acid, into which chemically
sugar might be entirely resolved, certain by-products

6

are always formed partly from the constituents of the sugar; viz., succinc and lactic acids, glycerine, propylic, amylic, butylic, and caproylic alcohols, and probably other substances still unknown (Watts' Dict.). Now, if the process is catalytic, why is the whole sugar not resolved into carbonic acid and alcohol? In the true catalytic formation of chlorine from hydrochloric acid by oxygen and copper salts, the whole chlorine is evolved. Besides all this there must always be furnished some albuminous matter, in addition to the sugar, as pabulum, and there is also produced cellulose and new protoplasm. Instead, therefore, of a more or less complex, merely chemical, change the yeast fermentation should be represented thus. A portion of yeast protoplasm converts into additional living matter like itself, some sugar with a trace of albuminous matter and salts. At the same time the surface layer dies and is converted into cellulose which remains as the cell wall, and the above products which are dissolved or volatilized away.

Likewise in the growth of bacteria we observe the same formation of by-products, whatever be the nature of the albuminous matters by whose destruction these creatures live. But what is more remarkable, Cohn found, and his observations have been confirmed by those of Mr. Dallinger and myself, that a specific odour of putrefaction is given out when they (bacteria and infusoria up to the rank of paramecia) are nourished without any albumen, *i.e.*, in a solution of tartrate of ammonia with mineral salts. This shows that in growth, the bacteria must form protoplasm out of the ammoniacal and other salts, and portions of it are then decomposed in vital acts.

Ultimately these products, when not dissipated as volatile, form nutriment for different generations and kinds of putrefactive infusoria, till finally nothing but a clear liquid is left, with a powdery sediment composed of unconsumable matters and bacterial and other germs. Thus in the course of successive generations is effected the reduction to simple compounds which is performed simultaneously by the myriads of plastids of different kinds which go to make up individuals of the higher orders.

Absorption. — In the absorption or retrograde metamorphosis of the tissues, we have, at first sight, a difficulty in the protoplasmic theory. But on closer inspection it proves more capable of explaining the phenomena than any other hypothesis. For if all the tissues are merely dead formed material and therefore passive, how are they to undergo that renewal which implies removal of old worn-out parts us well us deposition of new? It is true that by solution, oxidation, exosmosis, and other chemical and physical processes, a certain amount of effete matters is got rid of, as is allowed by Dr. Beale also, yet that does not nearly account for the bulk of the change in solid tissue which is supposed to be continually going on. In most works on physiology, it is tacitly assumed that although now the protoplasm may be the chief agent, embryologically speaking, for the production of the tissues, yet when once formed they possess vitality, and are renewed, particle for particle, by strictly vital processes as they are worn out, although no distinct meaning is ever given to the expression wear and tear as applied to living parts.

On the protoplasmic theory such a process is impossible. The immediate means is simply a kind of reversal of the essential function of the protoplasm, viz., consumption and renewal of itself from pabulum. In the case of one of the simplest forms of life, viz., the spore of mildew, the formed material may accumulate in the form of a thick membrane while the protoplasm wastes to a mere speck which retains its vitality in a dormant state until

"External conditions become again favourable, when the trace of living germinal matter soon increases, spreads through the thickened membrane, much of which it even consumes as pabulum, and the rapid growth already referred to is resumed. This rapid increase of germinal matter under favourable conditions is a fact of the greatest interest and importance in reference to certain changes occurring in disease of the higher tissues of plants, animals, and man. For we shall find that just as the germinal matter of the fungus may grow and live and produce new germinal matter at the expense of the formed material already formed, so the germinal matter of a cell of the highest organism may increase and consume its formed material. In this way we shall see that firm and scarcely changing tissue may become the seat of active change and ultimately be removed. Thus is the fatty matter of adipose tissue removed, and the hard compact tissue of bone scooped out to make room for new osseous texture. In this way the abscess and ulcer commence, and the 'softening' of cartilage and other hard textures is brought about. The pathological process known as 'inflammation' is due to the increase of germinal matter. In certain forms of cancer the process is seen in its most active, and to us most painful, form; for as the growth proceeds, not only is the formed material of adjacent textures rapidly consumed, but no sooner has the soft cancer texture been produced than it in its turn is consumed by new cancer-tissue, and this by more, until an enormous mass of soft, evanescent, spongy texture results, which destroys the poor patient by its enormous exactions upon his terribly exhausted system " (" Oxford Lectures," p. 579).

Again, speaking of cartilage, Dr. Beale says: " The germinal matter may even appropriate the formed material itself as we found appeared in the case of the formed material of mildew, epithelium, and other kinds of this substance " (" Oxford Lectures ").

Again, in speaking of the absorption of fat he says:

" As fatty matter is formed from bioplasm, so its removal is effected only through the instrumentality of this living matter. It cannot be removed until it has been again taken up and reconverted into bioplasm. Moreover the same bio-plasm is instrumental in both operations. In the one case taking certain constituents *from* the blood, increasing at their expense and then undergoing conversion into fatty and other matters. In the other, growing at the expense of the fatty matter already produced, and becoming resolved into substances which find their way back again into the blood, and which are at length appropriated in part by other forms of bioplasm of the body. . . . . In the winter, when the fat of the fat bodies of the frog are being absorbed, the bioplasm of each vesicle can be seen spreading around the fatty matter, which gradually diminishes in amount in consequence of its conversion into bioplasm. On the distal side of the vesicle, phenomena of another kind are proceeding. The bioplasm is there undergoing change and becoming resolved into substances which are imme-diately taken up by the bioplasm of the blood and the blood-vessels " (" Biopl." p. 138).

Dr. Beale sums up the action of the protoplasm thus : " Such is the marvellous power of this living material that there are probably few things in nature that are proof against its de-stroying power. The hardest material, even flint itself, yields to the slow but sure disintegrating action of bioplasm. The most insoluble materials as well as the most soluble are appro-priated " (p. 160).

In these examples we see how absorption is just as much a part of the function of protoplasm as deposition, and in fact it is in reality a mode of *growth* which is the distinguishing attribute of living matter. But these examples hardly do justice to the theory as a whole, for they apply either to disease or to the lower forms of individuality in which all the essential functions of life are concentrated in simple plastids. To see how removal of tissue forms part of an orderly purposive process just as much as deposition, we must go back to the laws of development. In all living matter the faculties of nutrition, function, and development are inextricably interwoven, and however one or other may be predominant at any period of life, still the others are never altogether absent. Schwann divided the faculties of the cell into plastic and meta-. bolic; the latter meaning those processes whereby chemical transformations are produced, and therefore principally concerned in secretion, which is the chief part of all functions, and the former including the formative and also germinal faculties. Thus the plastic, the metabolic, and the germinal may comprise the faculties of living matter. Beale looks upon the germinal as the fundamental one, and as lasting more or less through life and including the rest, he gives that name to the living matter. He points out that the same mass of protoplasm in the cells of secreting organs gives origin to two very different products, viz., the cell wall and the secretion, two things whose composition and properties are totally different, thus representing the plastic and metabolic faculties. The ultimate power of forming the tissues is only reached by a long series of descendence of one bioplast from another, and the germinal, or developmental faculty, not only survives in the adults of the higher individuals in ordinary nutrition, but "in fully-formed organs there exists a certain proportion of embryonic germinal matter which may undergo development at a future period of life, and if the greater part of this becomes fully-formed tissue, still there remains embryonic matter for development at a still later period, and so on" ("Croonian Lecture," p. 263). He instances the white corpuscles as examples of undifferentiated protoplasm which are descended from an early embryonic period and have more capacity for differentiation. To these he ascribes the power of forming cuticle and white fibrous tissue in the healing of wounds.

Beale's theory, in fact, harmonizes with, and realizes, in a more precise form, the theory of epigenesis of Caspar Wolff, who, we are reminded by Huxley and Häckel, is the true founder of the modern theories of development. Like Wolff, Beale traces the evolution of the whole complex organism of plants and animals to the growth, sub-division, and differentiation of a little mass of clear, viscous, structureless matter. The developmental history of living organisms is usually spoken of as synonymous with embryology, although that word properly applies only to the organism within the envelopes of the ovum. According to Häckel, the more correct term would be "ontogenesis."* Each individual of the higher orders, in the process of ontogenesis, passes through the lower stages which represent the permanent form of beings of lower orders of individuality. The process of their evolution or development, as a whole, consists therefore of building up, metamorphosis, and regression or decay. By many the word evolution is restricted

* "Gen. Morph.," vol. i. p. 53.

to the growth and differentiation up to maturity, while the
regressive metamorphosis is spoken of as something different,
and opposed to evolution. But Häckel will not allow this,
and regards the whole three as parts of one connected whole ;
for the regressive metamorphosis of some parts is so intimately
connected with the progressive development of others, that no
line can be drawn between them.*   Now, absorption is an
essential part of the process of metamorphosis, and although in
the higher orders of organisms most of these metamorphoses
and absorptions of temporary parts and organs take place in
the embryo, still, even in the adult, the tendency survives in
the same degree as the germinal faculty, as is revealed of both
in the phenomena of the healing of wounds, and of disease.
The persistence of the germinal faculty is also shown by the
phenomena of morbid new growths, and also by that of degeneration of the healthy bioplasts which forms such a large
element in the causation of inflammation, fever, and the contagious diseases. In all these the developmental faculty is
degraded to a lower level, while the metabolic and, at times,
the plastic faculties are often enormously increased, and there
is a corresponding consumption of albuminous pabulum, and
excretion of urea and carbonic acid. In the highest kind of
protoplasm of all — viz., the brain-bioplasts, the germinal
faculty must persist throughout the whole life, if, as is most
probable, we are to ascribe memory to it.

Beale, indeed, hardly does justice to his own
theory of absorption being the function of the protoplasm alone in the scanty observations directly
addressed to the question ; but it comes out more
clearly incidentally in his theory of the connective,
bony, and other tissues.

Instead of the important and numerous uses rather
fancifully attributed by Virchow and others to the
connective tissue, Beale thinks it of subordinate im-

* "Gen. Morph.," vol. ii. p. 18.

portance, and substantially refers it to imperfectly-absorbed remains of embryonic and previously-formed tissues. "Every adult organ may be said to contain skeletons of organs which were active at an earlier period of life" ("Biopl.," 105). The function of absorption thus belongs to the metamorphic part of the developmental process, and the proof of this is given by the fact that the connective tissue is almost completely absent in insects, in which, as is well known, the metamorphic process is most complete. Here the organs and textures of the larva are entirely removed by the masses of germinal matter which are destined to form the imago. These masses "absorb, remove, and, in fact, live at the expense of the tissue which is to disappear; and whether this change occurs physiologically or [in other organisms] pathologically, the process is essentially of the same nature" ("Croonian Lecture," p. 265). In a similar way, temporary muscular fibres are first produced and absorbed; temporary kidneys and other temporary organs of embryonic life illustrate the same process. In disease this absorptive faculty may be exalted and the reparative one destroyed. "In those changes which lead to the formation of pus, the removal of every texture is as perfect as during the pupa state of an insect, but the bioplasm constituting the pus corpuscles has no power to give rise to that which will take part in the development of new tissues" ("Biopl.," p. 106). I may conclude this subject with the absorption of bone, which illustrates, in a striking way, the function of the protoplasm in this process.

In the removal of the temporary bones in the fœtus the bioplasts in the central part increase and erode the spicula of temporary bone. The surface is eroded and scooped out into little pits by the bioplasts which consume the bony matter, and at last a cavity is formed which is occupied by multitudes of bioplasts, the descendants of those of the temporary carti- lage. It was discovered by Tomes and De Morgan that, even in the adult, the Haversian systems of the bones were in a continual state of change from absorption of old and deposition of new bone. The process is thus described by Dr. Beale :— Around the vessel occupying the Haversian canal, in the slight interval between the vessel and the surface of the bone, are seen a number of little bioplasts. These "little particles of bioplasm grow and multiply in the space in which they lie. The walls of this space (lacuna) are eaten away, and the lacuna becomes enlarged. As the hard material disappears, instead of a lacuna occupied by a single bioplast, we find a greatly- enlarged space, a gigantic lacuna, containing several bioplasts. One of these I figured as early as 1861. · The bioplasts of the adjacent lacuna increase in the same manner, and, by degrees, lamina after lamina of the osseous tissue of the Haversian rod disappears, and, in place of hard bone, we find soft, pulpy growing bioplasm occupying what is now [in the dead, dry bone] the 'Haversian space.' . . . The process of disinte- gration gives place to a very different operation. Of the bioplasts in contact with the bone, some, no doubt, die and disappear, others are, however, concerned in the production of soft formed material, which gradually becomes infiltrated with calcareous matter, as already described, and a layer or lamina of new bone results" ("Biopl.," p. 157). In this manner, without the de- velopment of any acid or chemically solvent fluid, this soft and pulpy bioplasm erodes and removes the hardest bony matter, such as the fangs of the temporary teeth. Those particles of protoplasm concerned in the absorption of bone can be shown by the carmine process.*

* In support of this view of absorption being a vital process, per- formed by the plastids or bioplasm masses, we may notice the forma- tion of the so-called giant cells in the absorption of bone. When the

On the Blood Formation.—Besides growth and absorption, the third general element of vital action in individuals of the higher orders is the preparation of the special pabulum, on the immediate conversion of which into living matter all continuance of life depends. On this subject I will only touch upon the points in which Beale's protoplasmic theory differs from the views previously held by physiologists. The elaboration of the blood to a state fit for pabulum to the higher tissues is performed by a series of protoplasmic masses of special, but lower, endowments than those of the higher tissues. While passing through these the crude nutriment is entirely decomposed, and

---

bone is exposed to irritation or pressure, as in the case of aneurism, or physiologically in the jaw during the formation of the tooth-sac, large, many-nucleated cells make their appearance. Wherever these are in contact with the bone a portion is eaten away, and what were before known as the lacunæ of Howship are formed. By this process of lacuna-formation the bone is gradually cleared away before the advancing tooth-sac, or other irritating cause which is producing its absorption. What is specially of interest to us in showing that, even in the living body, the true bone substance is actually dead, is the fact noted by Kölliker that the ivory pegs driven into the bones in the living subject in Dieffenbach's operation for pseudo-arthrosis were found to be covered with the typical lacunæ of Howship filled with the many-nucleated giant cells. And, besides, in many places human bone-substance was deposited on the eroded surface of the animal ivory. This corresponds to what is observed in the growth of the tooth-sac; for after the irruption of the tooth, the giant cells disappear, and the osteoplasts, or formative bone protoplasm masses, again clothe the surface of bone, whose formation, instead of absorption, again proceeds. This leads us to the origin of these cells, and we find that Kölliker considers they belong to the same category as the osteoplasts or formative cells of bone, and proceed directly from them; that there is a transformation of the one into the other by division; and that they may be seen in transition stages. Dr. Morison has also seen intermediate forms : " and also osteoclasts (giant cells) lying in close apposition on the one hand to the border of absorbing bone, and on the other to a twig of capillary vessel." See "Bone Absorption by means of Giant Cells," by Dr. A. Morison, "Edinburgh Med. Journal," Sept. 1873.

recombined into other compounds, differing often widely from it.   Thus, after the food is mixed with and dissolved by the special digestive juices, it is taken up by the bioplasts of the villi of the intestinal mucous membrane.   This taking up is not a mere physical absorption, nor a chemical process the like of anything known in the laboratory, but is strictly, like all other vital actions, a growth and death of portions of these bioplasts—the part that grows and lives constituting the white corpuscles which pass into the lacteals and finally into the blood-vessels, while the part that dies constitutes the serum, or true pabulum, which is still further elaborated by the bioplasts of the blood-vessels and those floating in the blood itself. " The food is not simply dissolved and caused to pass into the blood, as would be inferred from the description usually given, but millions of masses of bioplasm live and grow, pass through certain stages, and die yielding up the products of their death to be taken up by other bioplasm—particles situated in the walls of the vessels and in the blood itself" ("Biopl.," 25). The true nutrient part of the blood is thus always merely dead pabulum ; but the blood, as a whole, contains floating particles of living matter besides those fixed in the walls of the vessels, and generally called nuclei.   Thus are reconciled the opposite opinions respecting the life of the blood which have been the subject of so much contention.   The blood as a whole does not touch the living matter of the tissues in health, but it is only the dead fluid nutritive part which transudes through the capillary walls.   If from disease the living particles pass through the capillary

walls, the effects are always evil, and often fatal. In
ordinary nutrition and secretion, the living matter imbedded in each tissue or organ, selects certain constituents, and the fluid deprived of these is again
taken back into the circulation. The important part
played by the bioplasts of the capillaries in these processes is thus stated by Dr. Beale :

"I believe that the bioplasts of the capillary vessels play a
far more important part in the changes of the body than has
hitherto been supposed. They are as intimately concerned in
the process of secretion and excretion as they are in the selection, preparation, and distribution of nutrient constituents. The
bioplasts of the capillaries of the lungs are the agents by which
certain animal matters are separated from the blood and transferred to the air in the pulmonary air cells, and it is probable
they are also concerned in facilitating the changes which take
place between the gaseous constituents of the air and blood.
In connection with the capillaries of all secreting organs the
bioplasts are numerous, and they select and remove certain
substances from the blood and transfer them in an altered form
to the secreting cells of the gland. They are in great number
upon the vessels of the villi of the small intestines, in some
cases being so very close together as to leave little membranous
structure between them. These bioplasts of the intestinal
capillaries receive the nutrient substances after they have been
already once modified by the bioplasm of the epithelial cells of
the villi, and transmit them in an altered form to the interior
of the capillary, where many of their constituents are at once
taken up by the bioplasts (white blood corpuscles and minute
particles of bioplasm) in the blood itself."

These remarks entirely harmonize with the views of
many of the older physiologists on the importance of
the parenchymatous or capillary tissue—which to them
was the ultimate anatomical element—and more especially with those of Fletcher, who placed all the vital

powers of the capillaries in the portions of structure-less, pulpy, living matter interwoven with them. The number and size of these nuclei or bioplasts of the capillaries, are shown to correspond with the important functions assigned to them by Dr. Beale ("Biopl.," 287). In particular, the absorption and restoration to the blood of the nutrient fluid, altered during nutrition and the products of breaking up of the tissues, was assigned by Fletcher to the capillary veins in virtue of the living matter in their walls. This is almost literally repeated by Dr. Beale, who describes and figures these minute veins as destitute of muscular fibres and studded with immense numbers of bioplasts, whose appearance and functions are thus stated:

"The observer will be astonished at the great number of oval bioplasts in the walls of the small veins, as well as in the capillaries near the veins. These bioplasts have not, I think, been figured or accurately described, nor has attention been drawn to the very important offices they probably fulfil in connection with physiological changes that are constantly going on as long as life lasts. It must be obvious that bioplasts distributed in such number as are those in the walls of the small veins, perform other functions besides taking part in the formation of the tissue of the vein. As I have already endeavoured to show, the activity of change in an organ or texture may be judged of by the number of bioplasts present in it. In veins the bioplasts are many times as numerous as would be required to produce the very small amount of tissue entering into the formation of their coats. The blood in these small veins undergoes important changes, just as it does in the capillaries, and the agents concerned are the bioplasts. In short, physiologically, the small veins may be considered as part of the capillary system, and concerned in nutrition and in the removal of products of disintegration resulting from changes in the tissue."

Such is the general bearing of the protoplasmic theory on the formation and renewal of the blood, and it would be out of place to go further into that large subject here.

CHAPTER VI.

CELL AND PROTOPLASM HISTORY SINCE 1860.

WE have seen that in 1850 Cohn came very close to the protoplasm theory ; Remak also gave the name of protoplasm to animal cell contents, and although no one generalized the theory completely before Beale, in 1860, still several were working in that direction, and by most of the German authors the credit of so doing is generally given to Max Schultze, in 1861. How far the claims of this otherwise highly-distinguished naturalist are to be put into competition with those of Beale, is best shown by his memoir, " Das Protoplasma der Rhizopoden, &c. Leipzig, 1863." As this may not be accessible to many, I give here an analysis of the parts bearing on this subject :—

In Reichert and Du Bois-Reymond's " Archiv.," 1861, Max Schultze expressed the opinion that the dispute respecting the nature of the muscular and connective tissue corpuscles would easily be settled if people would give up some part of the prevailing views on the essential constituents of the cell, more especially in respect to the relation of the cell membrane and inter-cellular substance to the cell contents. At the same time, he said that the portion of the cell contents corresponding to the protoplasm of Von Mohl ought to be credited with a much higher importance, not only for the cell life, but also for the formation of tissues in the animal organism, than it had hitherto been. He rested these opinions on his own observations, and chiefly on the formation of the contractile substance of the muscular fibres of frogs and salamanders from the protoplasm of the embryonal muscle cell. Then, touching on Beale, he says—"No one who has the development of the general doctrine of tissue formation at heart will deny that the first requisite of progress in this direction is an accurate knowledge of the individual constituents of the cells, and their part in the development of the tissues. The need of such progress is even felt also in those quarters in which the cell theory has not yet been understood, and among such is to be reckoned the work of Beale" [on the "Simple Tissues," 1861], "which will not meet with the attention it otherwise deserves, because it stands outside the cell theory. Beale's 'Germinal Matter' is, it is true, essentially that which we call protoplasm, including also, certainly, the nucleus; and the 'formed matter,' that which is formed and gives form to the tissues is brought into an essentially correct dependence on the protoplasm. But of cells as elementary parts, or elementary organisms, of nuclei which are distinct from, and yet so necessary to the protoplasm, there is no word. For him the great and inalienable [unveräusserliche] discovery of the cell is of only historical interest" (p. 3). He then proceeds to state that he gives the protoplasm the same importance in animals as, since Mohl, the botanists have long assigned to it in vegetables. But the membrane is not thereby to be thought of no account, for it is necessary to the formation of rigid structures, for which purpose the protoplasm is unfitted, from its physical consistence. Although he speaks of the protoplasm as "properly speaking, the living substance of the cell," he does not clearly state that the membrane and the tissues are not living, and gives no clear notion of the relation of the cell membrane to tissues. He is still taken up with the demonstration that animal cells, or protoplasm corpuscles, can exist without a cell membrane, instancing the cells of the Hydra, the Amœba, the Myxomycetæ, &c.

7

This need not detain us, being now superfluous. He agrees with Priugsheim that the supposed primordial utricle, as a separate membrane, has no existence in the vegetable cell, and that the outer layer is merely protoplasm, as we saw Naegeli stated long before. He gives his own observations showing there is a similar outer layer of protoplasm in the animal cell which cannot be called a membrane, even when no distinct separation between the protoplasm layer and intracellular fluid takes place, as in fact it does more seldom than in the plant cell. In short, he considers established the fundamental analogy of the plant and animal cell as regards the protoplasm. He considers that the cell may be defined as consisting of a little mass of protoplasm and a nucleus, and that these constitute a single whole, and that the nucleus plays an important part, although he confesses what that may be is not as yet sufficiently known. He shows, also, that the movements of the pseudopodia and the granules are produced by active contractile movements of the protoplasm, as had been hypothetically supposed by Unger. In 1860 he had already used these words :—" I finally proposed to banish entirely the word Sarcode, which was repugnant as standing to a certain extent in opposition to the cell theory, and to substitute for it the word protoplasm, in which is expressed the triumph of the cell theory over these lowest organisms also,"—i.e., the Rhizopoda, which naturalists had found difficult to bring under the cell theory (with good reason, as we think still). In the essay of 1861 he says—" The opinion may be defended that the formation of a chemically different membrane on the surface of the protoplasm is a sign of commencing retrograde metamorphosis, and that the cell membrane belongs so little to the conception of a cell, that it may even be looked on as a sign of approaching decrepitude, or at least of a stage in which the cell has already suffered a considerable loss of its original vital activity." This was violently attacked by his German critics, and now (1863), after reading Beale, he does not re-affirm it with increased decision, but, on the contrary, draws back and apologizes for it as an opinion which one might defend as a whim (aus Laune). M. Schultze concludes his book in the following words—" I must once more expressly point out that Reichert's fears that the bases (Grundvesten) of the cell theory are shaken by my view of the cell are completely groundless. Nobody can be more deeply penetrated with the conviction than I that the doctrine of the cell, as the fundamental element of all animal tissues, is for all time inalienably assured. Far from wishing to put anything new in the place of the cell theory, I seek rather by my view of the body of the Rhizopod to bring its substance—the so-called Sarcode, hitherto standing outside the cell theory—under this theory. And, as regards my position towards Schwann's doctrine, I think that in many points we shall have to return to the purer form of the same. My observations force me ever more to the conviction that 'the correspondence in the structure and growth of animals and plants,' as Schwann entitles the scope of his celebrated researches, is much greater than people are nowadays inclined to believe" (p. 63). I cannot but think that the protoplasmic theory of Beale is a more simple and natural fulfilment of Schwann's object than Max Schultze's attempt to uphold the cell theory in words, while explaining it away to nothing in fact. And it is difficult to see what hinders him from following Beale entirely. But that he does not do so, and that he rejects the distinct declaration that all structure is dead, while the structureless protoplasm alone is alive, is plain from the above quotations. After this there can be no question of priority, for no one, then, was so near Beale as this author.

Reichert still after this continued to uphold the typical cell, and contested the fact of the internal movements of the sarcode of the Rhizopods; but his arguments are not deserving of any attention now, although Häckel, in his memoir on the Monera, 1869, took the trouble to refute them. Häckel claims for himself the merit of having done much to establish the protoplasm theory, in his " Monographie der

Radiolarien," in 1862, and other works, and summed up in his " Generelle Morphologie," 1866.

" The decisive and irrefragable proof that certain cells are destitute of all traces of a membrane, and consist solely of a little lump of semi-fluid mucilaginous cell-stuff (protoplasma), surrounding a nucleus, was first given by me ; in that I observed the penetration of solid particles into the substance of the protoplasm, and their accumulation round the nucleus. This was confirmed by the simple experiment of causing the amœba-like blood cells of invertebrates (Mollusca and Crustacea) to take up pigment particles into their interior, by means of their amœba-like movements and changes of form" (vol. i., p. 271). Having thus, like others, abandoned the cell wall, he still clings to the nucleus, to which he attributes important powers. But, in answer to Brücke, who insisted that many elementary organisms existed which were destitute of nucleus, he is compelled to admit the fact. And, in addition to the nucleated cell, he admits another kind of elementary organism, which he calls a cytod, or cell-like body, and which, in fact, is nothing but a little mass of protoplasm. In 1866 he divides the plastids, or morphological units of the first order, into cells—i.e., protoplasm with nucleus, and cytods, the same without nucleus ; while both of these may exist with or without a cell membrane. A mere lump of protoplasm is thus a naked cytod, or *Gymnocytod*, under which head are classed the Protogenes, the Protomœba, many monads, vibrios, &c. Under the head of *Lepocytods*, or protoplasm masses with a membrane, but no nucleus, are classed many Protista, such as certain *Rhizopods*, many *Algæ*, the spores of *Aphides, daphnidæ*, &c. To the Gymnocyta, or naked cells—i.e., protoplasm and nucleus—belong many ova, the partition-products of the same, many nerve cells, connective tissue cells, the escaped swarm spores of many *Algæ*, &c. To the Lepocyta, or the original cell with nucleus, protoplasm and cell wall, belong most plant cells and many animal cells. It is evident that, however valuable these precise divisions may be as questions of natural history, the essentiality of anything of even a twofold structure in the living unit is given up, and the existence of living masses, without any nucleus, is admitted by Häckel, as it was by Naegeli, A. Braun, Cohn, Brücke, and many others, before and since. He holds the plasma to be the "active material substratum of life, and which thus in a certain sense may be designated the 'life-stuff,' or, in a stricter sense, the 'living matter'" (p. 275). As to the nucleus, he expresses himself very doubtfully, and wishes to associate it with the formative power of the protoplasm ; then he says, in commenting on Beale's theory—" Certainly the nucleus, as regards its origin, is to be looked upon as a differentiation-product of the protoplasm, but in the sense that now the plasma and nucleus stand beside each other as co-ordinated parts, as to a certain extent different organs of equal rank, and which perform different functions" (p. 287). Then he follows much the same theory as Virchow, and assigns to the nucleus the power of propagation and of inheritance of hereditary character, while to the plasma belong the nutrition and adaptation to outward circumstances. While " in the cytods, in which the nucleus and plasma are not yet differentiated, we have to regard the whole plasma as the common organ of both functions" (p. 288). This, of course, gives up the idea of three, or even two distinct parts, as essential to the vital unit, which is the characteristic of the cell theory. The question is thus reduced to the use of the nucleus as an organ in certain physiological individuals of low order. (See p. 58.)

On this much-vexed question of the nucleus we have seen that Leydig, in 1857, while he rejected the cell wall, still held that the most important part of the cell was the nucleus, which, as it were, animated the protoplasm surrounding it.   Brücke, on the other hand, before Häckel, states that "the constancy of the appearance of the nucleus is subject to essential limitations if, as cannot be avoided, the cells of cryptogams are also considered, and it is not assumed that the nucleus must be present even where it is invisible"

("Die Elementar Organismen"). Likewise Max Schultze,
in 1854, and Kühue, in 1864, showed the existence of
organisms without nucleus.   Lockhart Clarke, in 1863,[*]
describes the formation of muscular fibre in much the
same way as Beale, the so-called nucleus being the
protoplasm itself.

" The nuclei concerned in the development of muscular fibres
have no envelope, or cell wall, in the proper sense of the word,
and these bodies are not entitled to be considered as nucleated
cells." . . . "However, there is little doubt that the muscular
substance is the result of some process carried on by the nuclei
themselves." . . . . The striated fibre, "instead of being the
product of a nucleated cell, would appear to be itself a kind of
cell *formation*, which at first finds its prototype in the organic
muscular fibre cell, and in which the investing sarcous substance represents the cell *wall*."

It is unnecessary to multiply examples of non-
nucleated organisms, as the fact of their existence is
now quite established ; and with respect to the position
of the nucleus question in the higher orders of living
beings, Stricker, in 1870, sums up the then existing
opinions by stating that its use is unknown, but that
it cannot be essential, because although when present
it divides on division of the cell, still other cells without any nucleus can live, divide, and propagate, so
" we must exclude the nucleus as an unnecessary factor
in the ideal type of an elementary organism" (Hand-
Book, Syd. Soc.).   Again, J. Ranke, in 1872, says—
" The nucleus arises out of the protoplasm, and always
lies embedded in it, and may become re-dissolved into
protoplasm; it contains essentially the same chemical

* " Qu. Microsc. J.," p. 8.

constituents as it, and is, when present, an important part of the protoplasm" ("Physiologie," p. 8). Van Beneden, in his recent memoir on the Gregarinæ ("Qu. Mic. J.," 1871), gives some general remarks on the cell theory, from which we may extract the following :—
" The existence of the monera, which have been the origin of all living beings, and whose simplicity is found again in the youngest Gregarinæ, proves the existence of the plasson [Protoplasm] (see note, p. 45), as the primitive condition. But in the plasson the nucleolus appears before the nuclear layer. If we identify plasson with the blastema, such as Schwann understood it, we shall return to the views of the celebrated histologist who assigned to the cell a centrifugal evolution." This is in accordance with what I have said at pp. 25, 27, and if we look upon the nucleolus and nucleus in the light of organs, in the lower independent individuals, as stated at p. 58, we shall escape the difficulties of this question. In one of the most recent notices of cell theories, viz., that by Professor Cleland,[*] substantially the same opinions are expressed, although the author raises a useless question whether certain non-nucleated organisms should not rather be regarded as cell nucleus, instead of protoplasm. From all this it is evident that general opinion is now in accord as respects the facts with Dr. Beale's statements on the nucleus in 1860, and I think his theoretical views of it (see p. 60) will appear to most to be in harmony with these facts. As regards the cell theory as a whole, also little difficulty will be felt in

[*] "Qu. Microsc. J.," July, 1873.

concluding that the general opinion now is that which
was first distinctly spoken out by Beale. Nevertheless, many still cling to the cell doctrine as a whole—
in name, at least, if not in reality. At which we can
hardly wonder, indeed, if we accept this description of
it by Ranke :—

"Of these component parts of the cell [cell wall,
protoplasm, and nucleus], one or other may be wanting
without the totality ceasing to be a cell. The nucleoli,
the cell wall, or the nucleus, may be wanting, and yet
we must designate the microscopic form a cell, or elementary organism." We have seen also that Häckel,
in spite of his admission of cytods, or rather by his
bestowing that name on bodies which have nothing of
cell form, shows the same unwillingness to give up the
name. Now, if any one choose to describe a gun-
barrel as a stockless gun without a lock, he is free to
do so; but what good purpose can it serve? Or is
there even any fun in it? The truth is, this clinging
to the mere name of the cell theory by the Germans
seems to arise from a kind of perverted idea of
patriotism and of *pietas* towards Schwann and
Schleiden. These feelings are no doubt commendable
and to be sympathized with to a certain extent, but
surely they are carried too far when we find Stricker
in 1870 taking no notice at all of Goodsir, and only
mentioning Beale once in the most cursory way.

While we find the protoplasm generally admitted to
be the simplest form of living matter, and credited as
the agent of much vital action, and even as the germinal substance from which all tissues proceed, yet it
is far otherwise with the doctrine that it is the *sole*

living matter in organisms with a complicated structure. Indeed in this Beale stands alone among living physiologists, just as Fletcher did nearly forty years ago. The doctrine has hardly even been properly criticized as yet; in fact, its significance has not been fully grasped, and people seem to be satisfied without further thought that a system which makes four-fifths of a man, including the muscles and nerve-cords, to be nothing but dead matter, must be an error of some kind. And this, with the erroneous ideas that it rests solely on the carmine staining process and requires the revival of the vital principle, further indisposes them to give the theory the attention it merits. Even those who have adopted Dr. Beale's anatomical ideas, and to a great extent those on the formation of tissues and secretions, hesitate to accept to the full the essentiality of the absolute and unfathomable distinction between dead and living matter, and that the latter must always be structureless and semi-fluid.

For instance, Dr. Carpenter expresses his general approval of Beale's doctrines, but in his edition of 1865 he still states that "new cells may originate in one of two principal modes; either directly from a previously existing cell, or by an entirely new process in the midst of an organizable blastema." Professor Tyson comments on this inconsistency, for it is obvious that this last mode is incompatible with Beale's theory. Professor Tyson himself, however, follows Carpenter in admitting that the formative power may reside in the germinal matter, but that we cannot, on that account, deny vitality to many kinds of formed material, inasmuch as they, e.g., the nerves and muscles, perform vital functions. It is obvious that both these commentators have either failed to appreciate the cardinal point in Beale's system, or they deny the truth of it. In the "Journal of Anatomy and Physiology" for 1867, the re-

viewer of Beale's "Todd and Bowman," while highly com-
mending the work done, and stating that Dr. Beale "alone
takes a place worthy of the British name beside the histologists
of Germany," still he is "unwilling to admit that the exterior
of the cell possesses none of the vital properties, and is a mere
passive, lifeless agent, and that indeed a large part—perhaps
the larger part—of our bodies, being composed of cell walls, is
in this condition." Dr. Bastian, in his "Beginnings of Life,"
objects to the absolute distinction into living matter and
formed material, stating that the simplest living things present
no such distinction of parts as those spoken of by Beale ; then
adds, "It has always appeared to me to be a very fundamental
objection to his theory that so many of the most characteristi-
cally vital phenomena should take place through the agency of
tissues—muscle and nerve, for instance—by far the greater
part of the bulk of which would, in accordance with Dr.
Beale's view, have to be considered as *dead* and inert " (155).
Again, "Dr. Beale's dictum that the matter which he calls
'formed material' is dead, we regard as a singularly founda-
tionless hypothesis, the maintenance of which is beset with
difficulties. If muscles and nerves perform work, such functional activity must be attended by tissue changes in their very
substance. How, then, is repair to be effected ? Not after the
fashion in which living tissues are renovated, for these, according to Dr. Beale, are dead, and therefore cannot be amenable
to the laws which govern the repair of living structures. I
have no faith, however, in the ability of carmine to discrimi-
nate the not-living from the living, and can only state my
total inability to accept the opinion of Dr. Beale." We do
not wonder that any one who could hold such a doctrine
as this should exhibit so much antagonism towards the evolution hypothesis. But how such marvellously abrupt tran-
sitions are brought about we are not told ; and Dr. Beale,
moreover, forgets to mention upon what evidence he feels himself entitled to make such positive and startling assertions.

To a certain extent, however, we find there is an agreement
between Dr. Beale's doctrine and that of other excellent observers. He says :*—"However much organisms and tissues

* Loc. cit., p. 48.

in their fully formed state may vary as regards the character, properties, and composition of the formed material, all were first in the condition of *clear, transparent, formless*, living matter." Surely, however, he is uttering something quite contra-dictory when he says, in effect, previously, and also in actual words subsequently :—" All that is essential to the cell or elementary part is *matter that is in the living state, germinal matter*, and matter that *has been in the living state—formed material.*" Such "formed material" as Dr. Beale speaks of may be necessary in order to support certain theories, but it does not actually exist in the simplest living things or elemental living parts—these are, as he himself has frequently stated, "perfectly structureless" (p. 156).

Dr. Bastian gives no reason for his want of faith in the carmine process, so his remark adds nothing to the subject. The truth is, the sting of Beale's theory to Bastian lies in the difficulty it offers to the supposed evolution of living beings through mere chemical action, which is the theme of Dr. Bastian's book. This subject will be touched on more at length in chapter xi., but in the meantime I would remind Dr. Bastian that the theory of life without a vital principle is not tenable except on the theory of a living matter anatomically one in a chemical state *sui generis*, for if the solid structures undergo an isomeric change at death, how do they retain their exact physical qualities?

Prof. Cleland (op. cit.) says : "In the present day the protoplasmic element has assumed an enormous importance, casting the nucleus into the shade ; while the reign of the cell walls has come to an end altogether. But to speak of life, as is sometimes done, as if it were an inherent property of a particular chemical substance, is surely going too far, and is a view which has nothing true in it which is not more than thirty years old ; for it has been long familiar to every one that life never exists without the presence of nitrogenous substance of an albumenoid character ; and though it has since been discovered that life in various instances exists in non-nucleated, structureless masses of protoplasm, that is a very different thing from life being a property of protoplasm" (253). He does not go deeper than this, and fails to tell us what life [vitality] is, if not a property of nitrogenous matter, nor how many kinds of nitrogenous matter may manifest vitality although not as a property. I presume, however, that the above may be taken to be an objection to the theory of the living matter being anatomically one. Why does he not say more clearly what he means?

CHAPTER VII.

THE application of the foregoing principles to the structure and functions of special tissues and organs must of necessity work a great change in our views of physiology and pathology. Much has already been done in this direction by Dr. Beale, although as yet almost single-handed, and to follow him into details would afford the most convincing proofs of the truth of the doctrine of the sole vitality of the protoplasm, which is still so little heeded. It would be out of place, were it not superfluous, for me to attempt that here, and I can merely refer all those who have the opportunity to study his own works, viz., the edition of "Todd and Bowman," "Bioplasm," the treatises on the diseases of the kidneys and the liver, and on disease germs, and numerous memoirs in the "Archives of Medicine," and the medical and microscopical journals. Nevertheless, for those who may not desire to enter on this study in all its completeness, I may give a full commentary on the theory as applied to two important tissues, viz., the muscular and nervous.

These, indeed, have been the chief stumbling-blocks in
the way of the acceptance of the protoplasmic theory,
and, singularly enough, even to those persons who
show the most inclination to reduce all vital phenomena to mere chemical and physical actions.

As the function of the muscles is intimately connected with that of the nerves, it will be more convenient to begin with the general question of the
structure and functions of the latter, although the full
discussion of the theory of the nerve-action cannot be
completed till we come to the muscles.   The anatomical element of the question is of primary importance,
for the differences between Dr. Beale and other anatomists on several essential points are irreconcilable,
and if he be right many prevailing theories will have
to be very much modified or abandoned altogether.   I
may, therefore, give an analysis of the chief features of
his anatomy of the nervous system and the physiological deductions founded upon it.

The development of the nerves always proceeds *pari passu*
with that of the tissue to which they belong : new nerve fibres
are never developed so as to influence old muscular fibres, nor
old nerve fibres caused to influence newly-developed muscular
tissue, but if wasting takes place the whole are removed
together, and if regeneration takes place the new tissue is
formed from formless spherical masses of germinal matter
complete ("Croonian Lecture," 1805—264).   The protoplasm
or bioplasm in many of the textures performs no other office
than the formation of them ; but, "in every part of the nervous system, however, more especially at the peripheral distribution and central origin of all nerves, active changes of the
most important kind are effected through the agency of the
bioplasm, and these continue throughout life.   Indeed, in
some instances, nerve action, which is dependent upon changes

in the bioplasm, never ceases for a single moment from birth to death " (" Biopl.," p. 167).

In many of the lower animals, *e.g.*, the common starfish, and some other members of the radiata, are seen very delicate fibres and masses of bioplasm, arranged so as to form an extensive network amongst the tissues, and the entire "nervous system" consists of such a network extending through all parts of the organism. Dr. Beale is convinced that an arrangement fundamentally similar in principle is common to all animals, from the highest to the lowest :—" The active part of every peripheral nerve apparatus is an uninterrupted network of extremely delicate fibres, which are structurally continuous with the masses of bioplasm in the nerve centres—these last, however, in the lowest classes being, as it were, so spread out as to render it difficult or impossible to define which part of the nervous system should be considered *peripheral* and which *central*" (" Biopl.," 181).

" The nerve fibres composing the nerve trunks, and those finer branches which unite to form dark-bordered nerve fibres, may be arranged in the following subdivisions, according to their distribution :—

"1. Nerve fibres passing towards a centre—*Afferent fibres.*

"2. Nerve fibres passing from a centre—*Efferent fibres.*

"3. Nerve fibres connecting nerve centres with one another —*Commissural central fibres.*

"4. Nerve fibres connecting the peripheral ramifications of nerves and peripheral nerve organs with one another—*Commissural peripheral fibres*" (" Biopl.," p. 182).

The first three of these subdivisions are in agreement with general opinion, but the last leads to the question whether the nerves terminate by free ends in other tissues, or in certain end-organs, or on the other hand constitute closed circuits by looped extremities. On this possibly hangs the whole theory of the nature of the *vis nervosa* as a current force, so his exposition must be given somewhat fully.

" The active part of the nerve fibre in adult vertebrata invariably consists of a very delicate compound thread, which exhibits a slightly fibrous character, and is composed of an

olco-albuminous material. Connected with the threads, at
varying intervals, are oval masses of bioplasm. In highly-
sensitive peripheral nerve organs, and in the motor nerves of
muscle, these masses of bioplasm are very numerous, and, in
some cases, are almost continuous with one another; but in
less sensitive textures the masses of bioplasm are often separated from one another by a distance of one-hundredth of an
inch or more. In all these cases these bioplasts, or 'nuclei,'
are situated very close together at an early period of development, and at first the tissue which represents nerve, consists of
bioplasm only. As the tissue advances towards maturity, the
masses of bioplasm become gradually separated from one
another by a greater extent of fibre; but at all periods of life,
and in all peripheral branches of nerves, these bodies are
present" ("Biopl.," p. 171).

"The active part of the nerve fibre distributed to the peripheral nerve organ which receives the impressions, exhibits the
same general structure and anatomical arrangement in all
cases. It is invariably a pale, very transparent, faintly granular, but, in the natural state, perfectly invisible cord, composed
of still finer fibres" ("Monthly Microscopical Journal," 1873,
p. 173).

These fine fibres are themselves compound, and "fibres
often pass off at an angle from these fine nerve fibres, and
divide and subdivide, joining others, so as to form a network,
the meshes of which vary very much in diameter in different
cases. Every one of these delicate fibres, of which some are
not more than one-thousandth of an inch in diameter, must
be regarded as composed of still finer fibres, which, after
leaving the branch under observation, pursue opposite directions. In using the term *network*, therefore, I do not mean to
imply that fine nerve fibres unite with each other after the
manner of capillaries, but merely that the *bundles* of fibres are
arranged like networks. The fibres composing the bundles do
not anastomose. In lace the appearance of such a network of
fibres is produced; but every apparent thread is composed of
several, each of which pursues a complicated course, and forms
but a very small portion of the boundary of any one single

space. Proceeding from the finest nerve fibres, no fibres exhibiting *ends*, or terminal extremities, can be detected, and the general conclusion to which we are led is, that nerves are arranged to form continuous strands of fibres, which pass amongst the elementary parts of the tissues, but neither become continuous with them, nor terminate in free extremities in or upon them" ("Biopl.," p. 172).

"In all cases, as far as I can ascertain, the ultimate terminal fibres are pale and granular, exhibiting nuclei at varying intervals, but are distributed upon precisely the same plan. I am of opinion, therefore, that there is not such a thing as a true end to any nerve fibre" (p. 173). The diameter of fibres composing the terminal plexus is often less than one-thousandth of an inch.

"Every fibre of this network is compound : so that, perhaps, the term 'plexus' more truly describes the arrangement. 'Plexiform network,' I think, expresses the character of the arrangement still more exactly.

"Some have said that my view accords with the old idea of loop-like terminations of nerves ; and this is, in the main, true, but the course of one single fibre forming the loop is far more extensive than was supposed by the older observers, and the looped fibres divide and subdivide into finer fibres. This diagram is intended to represent a plan of the arrangement which is shown to exist in many tissues according to my observation" ("Croon. Lect.," p. 237).

These views are not generally admitted, especially by the German histologists who follow, for the most part, Kühne, who holds to the existence of terminal extremities in several motor and sensitive organs, and Pflüger, who describes the termination of efferent nerves in the protoplasm of secreting organs. Yet, after reviewing all that has been advanced against him up to 1872, Dr. Beale still persists as follows :—

"I am quite convinced that numerous specimens I have made fully justify me in maintaining the general proposition that, in all cases, the terminal distribution of nerves is a plexus network, or loop, and hence that in connection with every terminal nervous apparatus there must be at least two

8

fibres ; *and that in all cases there exist complete circuits, into the formation of which central nerve cells, peripheral nerve cells, and nerve fibres enter.* All these elements are in structural connection with each other " (" Biopl.," p. 174).

It would be superfluous to go into the details of the controversy, as Dr. Beale has reviewed it so recently, including the elaborate memoir of Dr. Klein,[*] but some points will be touched upon in discussing the theory of muscle and nerve action.

On the Nerve Trunks.—"Every peripheral nerve network is connected with its nerve centre by fibres, and whenever the distance between the centre and peripheral organ is considerable, the nerve fibres are protected from each other, and from the tissues through which they pass, by a thick layer of oleo-albuminous matter, which forms an investment to each bundle of delicate fibres, by which it is insulated and separated from its neighbours, and from other structures, by a distance equal to from five to twenty times its own diameter. When the trunk passes through narrow canals, as through holes in the cranium, this insulating protective covering is much reduced in thickness, so that a large bundle of nerve fibres is made to pass through a space not more than one-fourth of the diameter which the nerve-trunk possesses in other parts of its course. The fibres which have this thick covering are known as 'dark-bordered fibres,' from the dark double contour line they always exhibit when examined in water or weak serum ; the covering itself is known as the ' white substance of Schwann,' or the 'medullary sheath' " (" Biopl.," p. 175).

It forms a tolerably uniform layer round the core, which it probably insulates and protects, like the gutta-percha does the copper wire of the marine cable. Also this white substance of Schwann is so constituted as to interfere with the free passage of fluid. It is not very permeable to aqueous or albuminous solutions, and thus a uniform degree of moisture of the axis

---

* " Monthly Microscopical Journal," vol. vii. p. 156. Dr. Beale's reply is at p. 253 of the same volume, and appears to me satisfactory in all the essential points.

cylinder is preserved ("Monthly Microscopical Journal," 1872).

"What appears as the single core or '*axis cylinder*' of a nerve fibre in the nerve trunk is *formed by the coalescence of very numerous fine fibres, each coming from a different central nerve cell.* In following a single dark-bordered, or other nerve, towards centre or periphery, we find that it divides and sub-divides into a great number of fibres, which pursue different and often opposite directions, one passing towards the centre, and the other to the periphery. And these are implanted in different parts of the nerve centre or peripheral organ at con-siderable distances from one another" ("Biopl.," p. 177).

The effect of the frequent crossing and interlacing and change of course of the nerve fibres, Dr. Beale pointed out, is to prevent the complete paralysis of either motion or sensation of any part by injuries of a moderate number of nerve fibres.

The so-called pale fibres of the sympathetic do not differ essen-tially from the dark bordered nerves. There is simply no medullary sheath, where the distance between the nerve centre and the peripheral distribution of the nerve is not very great. "Where, however, the ganglia, or peripheral organs are connected with nerve centres at a considerable distance off, a number of fibres having this investment are found ; so that amongst the sympathetic nerve fibres we find dark bordered nerve fibres. In the bladder of the frog I have observed that when the distance between the ganglia and the peripheral dis-tribution of the nerve fibres is considerable, the fibres have the dark bordered character, while on the other hand, if the peri-pheral distribution is near the ganglion, the ultimate nerve fibres are connected with the latter by pale fibres only" ("Biopl.," 176). Dr. Beale has also demonstrated the existence of very fine fibres running close to the dark bordered ; these fibres in fact result from division of the dark bordered fibres, and are, in fact, a continuation of them near the point of their distri-bution.

On the Nerve Centres.—While in the lowest animals there is no obvious distinction between peripheral and central parts of

the nervous system, in the higher invertebrata and in the vertebrata it is probable that the nerve tissue collected in the nerve
centres exceeds in amount that spread out amongst the tissues
in all the other parts of the organism.

"Each *central nerve cell* consists of a mass of bioplasm surrounded by formed material, which last is drawn off at *two or
more points* into fine threads. These divide and subdivide into
still finer ones at a short distance from the cell, and are, in fact,
processes of the nerve cell which become nerve *fibres*." The
processes invariably take opposite directions soon after they
have left the "cell." In vertebrata there are two principal
kinds of central nerve cells which are very distinct from one
another, and probably differ in function not less than they do
in structure. These are,

1. *The angular, or caudate nerve cells.*

2. *The oval, pyriform, or spherical nerve cells* ("Biopl.," p. 186).

The latter alone are considered to be centres of evolution of
nerve force, while the caudate cells are merely commissural
organs. Dr. Beale denies the existence of apolar or unipolar
cells, and maintains that there are at least two fibres to every
nerve cell.

The spherical, oval, and pyriform nerve cells are composed
of masses of protoplasm, from which two or more fibres appear
as if drawn out, and are curved and coiled as if they continued
to grow, or were spun off as the cell revolved. These cells are
found in all the ganglia of the sympathetic, and on the posterior
roots of the spinal nerves, the gasserian ganglia, &c. In some
ganglia cells there is a straight and a spiral fibre. A singular
ganglion cell has been described by J. Arnold, but several
German authors have doubted the authenticity of Beale's pyriform ganglia, the drawing of which is copied in most text-
books. The actual specimen has, however, been shown to
many observers under the 1-25 (1800 diam.). Among the rest
I have seen it, and can testify to the correctness of the drawing,
and to the impossibility of such an appearance being produced
by any accidental tearing out of connective tissue or any such
cause. Dr. Beale concludes that these spherical, oval, and pyriform cells are the sources of nervous power, while the caudate

cells are more probably concerned in the radiation of the nerve currents.

" From the cells of the sympathetic ganglia of man and vertebrata several fibres proceed, and pass in different directions soon after they leave the cell. Bundles consisting of fibres from many different cells leave the ganglion from different parts of its surface, and pass by circuitous routes towards their destination, each bundle being composed of fibres from many different cells situated in different parts of the ganglion. Ganglia are extremely numerous in the sub-mncous tissue of the alimentary canal of, all mammalia, and in the human subject multitudes may be demonstrated at short distances from one another. Connected with the nerves in the pelvis of the kidney I have also demonstrated numerous ganglia of the same kind. From every one of these, bundles of nerve fibres pass to be distributed to the cortex of the organ. The fine nerve fibres of the kidney are distributed to vessels, and also to the uriniferous tubes " (" Biopl.," 193). " The early development of the spherical and oval cells, and their large size at a time when the caudate nerve cells are not to be distinguished, their constant presence, their growth, and multiplication in the adult, and probably at an advanced age, and their peculiar structure—at least in some animals—their situation, as regards the nerves to which they belong ; and especially the fact that these are the only cells constituting the nerve centres upon which the rhythmic contraction of detached portions of the cardiac muscular tissue depends, have led me to look upon them as the *sources* of nerve power " (" Biopl.," 196).

The Angular, or Caudate Cells.—These are characteristic of the great central nerve organs of vertebrata, the brain, and spinal cord, and attain their maximum of development in man. The peculiarity of these cells is, that lines are seen traversing them, from each of the many fibres connected with them, and passing to every other fibre. Fibres from different caudate cells unite to form single nerve fibres. " In passing towards the periphery these compound fibres divide and sub-divide, the resulting sub-divisions proceeding to different destinations. The fine fibres resulting from the sub-division of one of the caudate processes of a nerve cell may help to form a vast number of dark bordered nerves, but it is, I think, certain that *no single process ever forms one entire axis cylinder*" (187).

"It is probable that the caudate nerve cells are not *sources* of nerve force. These cells are fewer in number, and comparatively insignificant in the lower vertebrata, particularly batrachia and fishes. In the invertebrata they do not exist at all, and it is doubtful if any 'cells' precisely corresponding to them are to be found in their stead. The bioplasm of the nerve cell is embedded in the material which exhibits the lines crossing in all directions, and no doubt this substance is formed from it ; but as far as I have been able to ascertain, no nerve fibre arises from, or is connected with, the bioplast (nucleus, or nucleolus). It appears probable that the caudate cells are the stations at which nerve fibres pursuing many different directions decussate and change their course" ("Biopl.," p. 190).

We have thus a central counterpart of the commissural peripheral plexus formation, and it is interesting to note that Professor Bain, looking at the subject from a totally different side, arrived at conclusions concerning the arrangement of the central nervous mechanism agreeing in all important particulars with those of Dr. Beale.

"From the foregoing observations the reader will be led to conclude that I regard a nervous apparatus as consisting essentially of fine fibres and masses of bioplasm, which form uninterrupted circuits. The fibres are continuous with the bioplasts, of which some are central, some peripheral, and grow from them. Currents emanating from bioplasts at one part of the circuit would influence the changes in the bioplasts in another part, and the last react upon the first" ("Biopl.," p. 209).

Such is a brief view of the anatomical elements of the nervous system according to Dr. Beale. The most important additions made by him to our knowledge are, the extreme number and minuteness of the fine fibres, and the compound nature of even these; the commissural nature of the caudate cells, and the peripheral termination of all nerves in loops. By means of such anatomical elements we can see how the most complex functions of the nervous system as a whole, viz., afferent, efferent, commissural, and possibly inhibitory, can be performed. And also we have the possibility of a current force acting by induction from a closed circuit shown for the first time if the loop form is proved true, and sufficient insulation be rendered probable.

By the foregoing, we see how the nerves may consist of merely dead formed material, if considered merely as constituting an apparatus for conducting a stimulus from one living part to another, and it is in this sense that the nerves are generally spoken of when no other qualifying word is used. But like other parts, even the conducting cords require the presence of their special protoplasm masses or bioplasts for their formation and repair. And besides, it is in these that the nerve current (whatever be its nature) originates and not in the fibrous matter itself. As proofs of this, Dr. Beale adduces the following facts.

"That these masses of germinal matter, which I have shown to be numerous in all ultimate nerve fibres of all nervous organs, besides taking part in the formation of the fibres, are concerned in nervous action, appears to be probable from the following facts :—

" 1. They are very numerous in the peripheral ramifications of all nerves.

" 2. All special peripheral nerve organs, as the retina, the expansion of the olfactory and auditory nerves, the papilla of touch and taste, as well as the peripheral nervous expansions beneath sensitive mucous membrane, the skin, &c., are re-

markable for the great number, as well as for the large size of
the masses of germinal matter.

"3. The proportion of germinal matter is always very great in
nerve centres which are the principal seats of development of
the nerve power.

"The principal change which takes place in a texture, which
in health appears to be but slightly sensitive, and becomes
eminently so when inflamed, as the peritoneum, is a very great
increase in the germinal matter which it contains, and this
often proceeds to such an extent that the ramifications of the
nerves appear as lines of oval masses of germinal matter, so
that when a tissue, which in the healthy state gives no evidence
of sensation, becomes acutely painful when inflamed, the feeling of pain must be due, in some way, to an increase of the
germinal matter of the nerves as well as that of other tissues.

"From a consideration of the facts, we are led to conclude
that the nerve fibre in all cases transmits the nerve current as
a conductor, and that pressure, &c., on any part of its course
will affect the rate of transmission of the current, and the conducting property of the fibre. The nerve current itself results
from changes occurring in the germinal matter or in the substances formed by it, and it is probable that the masses of
germinal matter in the peripheral nerve organs may give origin
to nerve currents as well as those in the nerve centres. In
disease, the currents formed at the periphery of the nerves
probably undergo an increase in intensity" (Beale, "Med.
Times," May 22, 1869).

"5. That where, as in the sensitive papilla upon the toe of the
frog, the nerve organ is more acutely sensitive (or more active
in any other way) at one part of the year than at others, its increased activity is associated with a great increase in the
amount of bioplasm" ("Biopl.," p. 206).

Thus, in short, the nerves may be looked on as a
system of mere dead conducting cords, studded at
short intervals with bioplasts or little masses of living
matter, which, besides their other living functions act

as little batteries from which the *vis nervosa*—a mere dead force like all other forces, and possibly electricity—is evolved. This view harmonizes with the facts which indicate living action in them and distinguish them from mere telegraph wires. Among these may be noted, excitement by stimuli, such as pinching, pricking, and other mechanical and chemical irritants in any part of their course when entire, and even in the peripheral ends when cut; the cumulative action of the *vis nervosa* shown by the greater effect of stimulation of a nerve at a distance from its muscle than near (Pflüger); the increase of velocity of nerve current as you approach the muscle (Munk); and a variety of other facts, showing that "nerve excitement is not simple conduction" (Ranke); for if the impression conducted were merely the propagation of an impulse, like waves it would grow weaker the farther from the point of excitement owing to resistances.* The influence of the electric current in exalting and lowering the irritability of the nerves is also, on the whole, probably in harmony with this view of the constitution of the nerve cord, for a weak galvanic current increases the irritability to other stimuli, while a strong current weakens and ultimately destroys it, just as we find with all other stimuli to vital action, and in fact, irritability is always used by Fletcher as synonymous with vitality. It is also in harmony with the

---

* The explanation of Pflüger is that the molecules of the nerve in succession disengage active energy, and each stimulates its successor, but the increase of action ("avalanche-like") shows that each molecule disengages more force than the one before. Beale's is evidently a much more natural explanation, for in it each little battery of protoplasm is set in action by the passing current, and contributes its quota to the current which is thus really swelled like an avalanche.

fact that the conducting power of the nerves is impaired or destroyed by injuries to their integrity of every kind ; such as cutting, even though the cut ends are at once replaced in contact, bruising, tying, burning, cauterizing, or chemical alteration, all which impair or destroy vital action, even though they would little, or at all, interfere with electric conductivity.

Whatever the exact nature of the vis nervosa may be, Beale thinks "it is at least not improbable that the varying effects noticed in connection with the nervous system may be determined by alterations in the intensity of the current and in the conducting properties of the fibres, instead of being due to the transmission of *different kinds of nerve force.*"

This is in harmony with the general opinion of physiologists. The motor and sensitive nerve fibres are similar in their fundamental physiological properties, and both are said to be equally capable of conducting the afferent or efferent nerve current, as was inferred by Lewes, and afterwards supported by the experiments of Kühne, Phillipeaux, Vulpian, and Rosenthal. Quite recently, however, Vulpian has repeated his experiment of soldering the lingual and hypoglossal nerves ; and he now attributes the motor phenomena to conduction through the chorda tympani, and not through the lingual (sensory) nerve. Both motor and sensitive nerves are likewise equally capable of transmitting the electric current in both directions, and both display the negative variation under excitement.

We must, however, remember that the specific irritability of nerves differs towards poisons ; the woorari, for example, acting only on the motor nerves, so we must infer specific differences of vital properties in different nerves although the nerve force may be of one kind.

The nerves of special sense were thought to convey directly an impression of the qualities of outward things to the sensorium by a specific sensation in them, but the fact of common or electrical irritations of the nerve trunks causing the sensation usually dependent on the natural peripheral irritation is against that. The optic nerves under the action of common irritants cause the sensation of light, but neither they nor the fibres of the retina itself are susceptible to the stimulus of light, except when they are connected with the rods and cones. The specificity of the nerve influence does not thus lie in any specific kind of excitement of the mere nerve cords, but must be referred to the central organs to which it is conducted. The mental organs, which are excited by the stimulus conveyed through the nerves of special sense, are so excited because, from the special constitution of their protoplasm, they are only competent to produce that determinate sensation. "The same stimulus when it affects different mental organs, will be interpreted according to their several specific energies" (Ranke, p. 693). If, however, we are wholly to deny the transmission of specific irritation by the nerves several pathological and therapeutical theories must be much modified, and the specific effects of poisons must be referred solely to absorption, except in so far as a mere plus and minus of action is spoken of in organs connected by nervous sympathy. But we must not forget that, even if the vis nervosa is a single physical force, it may be "capable of an almost infinite number of variations or gradations," like the shades of colour or tones of sound ; and without such differences how can we explain the differences of impression of sweet and bitter conveyed by the same nerve, for example, or any specific impression at all conveyed by them ? so that Fletcher's theory of the conveyance of specific irritations by his respiratory system of nerves of organic sympathy (corresponding to the central vaso-motor system now spoken of) may still be tenable. He considered that the stimulus of poisons, besides being diffused by absorption into the blood, might be conveyed rapidly by means of the common centre in the spinal cord through all the nerves of organic sympathy, and only take effect on parts which were susceptible of its influence.

On the other hand, we must also remember that there must be a specific apparatus at the periphery of sensific nerves which shall enable them to take cognizance of, and express as a stimulus in *vis nervosa*, the different forms of pressure-force into which the objective cause of the special senses, except taste and smell, may be resolved. Helmholtz has revived the theory of Young, that distinct peripheral nerve apparatus in the retina take cognizance of the cardinal elements of colour ; and has extended the same theory to the sense of hearing, in which he supposes that a separate nerve fibre takes cognizance of each definite note to which the rods of Corti vibrate in consonance. The structure of the retina in birds and reptiles, as described by Max Schultze, and the extreme fineness of the ultimate peripheral nerve fibrils, as described by Beale, give an anatomical basis for the hypothesis. Thus the nerve force, as a stimulus, may be very similar everywhere, and may be a mere sign, whose specificity lies in the interpretation given at the nerve centre where it acts ; and also in the speciality of the peripheral apparatus, enabling the living matter to take cognizance at all of the different forms of external force as stimuli. Nevertheless, we can hardly suppose the existence of separate nerve fibres for all the numerous varieties of sensations depending on stimulation through nerve fibres ; so we must conclude there are some qualitative differences in the vis nervosa capable of being conveyed through one and the same fibre.

Beale would apparently reduce the whole of efferent nerve action to causing a simple plus or minus of muscular contraction, voluntary and involuntary, and the greater part of the afferent nerve action to furnishing the stimulus for the latter; for he denies all direct nerve action on the living matter in secretion and nutrition, any such action being merely indirect, through the change of calibre of the capillary arteries from muscular action. His theory is here surrounded with difficulties, and can hardly as far as I can see, be reconciled with the numerous facts, which show, not only mere quantitative alteration of secretion and nutrition, from altered supply of blood produced by efferent nerve irritation, but also alteration of quality.

The fact of the influence of motor nerves on glandular secretions is now universally recognized, and the flow of saliva from stimulating the distal end of the cut chorda tympani has become a stock experiment in the class-room. Beale refers this solely to the increased supply of blood to the secreting protoplasm masses caused by the influence of the nerves on the capillaries. But against this view is urged the fact that stimulation of the sympathetic nerve fibres supplying the gland, likewise increases—though not to the same extent—the flow of saliva, although at the same time it causes contraction of the capillary arteries with a diminished flow of blood through the capillaries; the quality of the secretion is at the same time altered, and it is rendered more viscous and richer in salivary corpuscles. It is, therefore, quite possible that the chorda saliva results from the action of nerves on the vessels, while that of the sympathetic, from action on the secreting protoplasm, as supposed by Pflüger and others. Besides the above, we may adduce Ludwig's experiment of exciting secretion in a gland cut out of the body by nerve stimulation, although no blood was present. To these I have seen no answer by Beale, but I believe his chief reason for dissenting from this view rests on the anatomy of the question; for, as we have seen, he denies that the nerve fibres ever terminate in the glandular protoplasm. Without questioning the

accuracy of Beale's anatomical view of the nerve termination, I can see no reason for doubting that
glandular protoplasm may be influenced directly by
the vis nervosa, even if the nerves are distributed
as he supposes, just as all protoplasm is susceptible to the stimulus of heat, electricity, and other
active forces, although we may not yet understand the
exact mode in which the force is transmitted. The
same remark applies to the direct action of nerves on
nutrition, under certain circumstances. Of this there
are too many well-ascertained examples* to be explained away by the argument of Dr. Beale—viz.,
that nutrition and growth take place independently
of all nervous action, and are, in fact, most active
before the nerves themselves are formed. Because
these operations take place under other stimuli, with
pabulum and conditions, than nerve stimulus, that is
no reason why nerve stimulus should not influence
them when developed; not, of course, in any sense
being the source of their vital power, but merely rousing
that into activity, like other stimuli. Dr. Beale seems
lately to have changed his opinion on this subject.

On the Nature of the Nerve Current.—The chief
theories of the nature of the vis nervosa are—first,
that it is a molecular force, like electricity; or, second,
that it is electricity itself; or, third, a chemical action,
propagated from particle to particle, liberating energy,
like a train of gunpowder set on fire by a spark at
one end. To this last, Dr. Beale objects that, if by

---

* See Laycock's papers in the "Medical Times," also H. Power's,
in the "Practitioner," 1873; also Ranke, "Physiologie," p. 83, and
many other sources.

that is meant a mere chemical slow combustion, such
as Liebig at first supposed to occur in the muscles, it
is quite untenable, as the nerve is a slow-growing
tissue, and is incapable of undergoing such changes,
and, as a matter of fact, is not so consumed. These
objections, however, would not apply, if by chemical
we understand those changes sui generis in the protoplasm, called by him and Fletcher vital; in fact, it is
a postulate of life, that changes involving consumption
and regeneration of protoplasm be constantly taking
place. It would, however, be necessary in that case
that the protoplasm should be continuous, by however
thin a thread, and always terminate in contact with
the protoplasm of the part to be influenced, because
vital action cannot be transferred to the smallest
distance. Beale's anatomy of the nerves, and with
their termination in loops, is fatal to this theory, if
vis nervosa influences any part with which the nerves
are not in continuous contact. The first theory—viz.,
that it is a specific or peculiar molecular force, easily
transformable into electricity and heat, but yet not
electricity, is treated by Dr. Beale with a degree of
scorn which I am at a loss to understand. He says it
is a very odd thing that people should have no difficulty
in referring the phenomena of the nerve current to a
specific mode of force correlated with heat and electricity, while they turn up their noses at the idea of
vitality depending on "some equally undiscovered
form of force having no connection with primary
energy or motion" ("Monthly Microscopical Journal,"
1872, p. 177). It is certainly very strange that he should
not see the vast difference between his revival of the

vital principle, and the assumption of a specific mode of molecular motion, in addition to those distinct forms of current force already—but not so very long—known, viz., electricity, galvanism, and magnetism. Thus, rejecting the other theories, Dr. Beale plainly adopts the proposition that the nerve force is simply electricity and nothing ·else. The arguments in favour of this idea have been long known, but Dr. Beale sums them up as follows :—

"The general arrangement of the fibres and cells in all central and peripheral nerve organs ; the structure and arrangement, with respect to one another, of all nerve fibres which pass through a considerable distance before they reach their destination ; the manner in which nerves act upon contractile tissues, and the circumstance that a current of electricity produces a similar action ; and the fact that electricity is set free in special organs which are very rich in nerves, but which do not differ in any essential particulars in ultimate arrangement from other nerve organs in which electricity is not set free, render it probable, as it seems to me, that the current transmitted by the axis cylinder is ordinary electricity, and that all the effects produced upon other tissues depend upon the transmission through nerve fibres of currents of electricity varying in intensity" ("Monthly Microscopical Journal," 1872, p. 176).

Bioplasm of the nerve tissue does not differ much from other bioplasm, and it is probable that "considering the character and arrangement of the bioplasm matter, and its relation to the formed material in all tissues, it is not unreasonable to conclude that currents, and perhaps of the same nature as those discharged by the nerve organs, are set free, but that it is only in the case of nerve that an arrangement exists suitable for insulating the currents, and for rendering evident variations in their intensity, rate of transmission, &c."

The arguments against the electrical theory of the vis nervosa are summed up by De Bois Reymond :—

"Every attempt to identify it with the electric current as it circulates in a telegraph-wire, must appear hopeless, even if a circuit, such as would be necessary for the supposed nerve current to circulate in, were anatomically demonstrated. Thus to the other arguments against this view of the nervous agent —that the resistance of the nerve tubes would be far too great for any battery to send an available current through them— that the physiological insulation of the nerve tubes from each other would be impossible to explain—that the effect of liga- ture, or of cutting the nerve and causing its ends to meet again, would be equally obscure. What we have termed the nervous agent, if we look upon its very small velocity, in all probability, is some internal motion—perhaps even some chemical change of the substance itself contained in the nerve tubes, spreading along the tubes, according to the speaker's experiments, both ways from any point where the equilibrium has been disturbed; being capable of an almost infinite number of variations or gradations, and of so peculiar a character as to require the unimpaired condition of the nervous structure" ("Matter and Force," B. Jones, p. 130).

To this Beale rejoins, in substance, that he has discovered the anatomical conditions of a circuit, although he cannot speak with equal confidence of insulation; that the experiments with strong electric currents through nerve-fibres, though undamaged, are unsatisfactory; that the loss of excitability, by death, without loss of conducting power of electricity, may depend on changes as to moisture and of currents of fluid, caused by the death of the protoplasm; and to· the objection of the uncommonly slow velocity of the nerve current compared to that of electricity, he says that it is not yet known how slowly electricity would travel through such a bad conductor as the axis cylinder, and certainly it would go immensely slower

9

than through a copper wire. But of the tremendous
loss of power incurred by sending electricity through
such a bad conductor, and to the fact that the vis
nervosa is not transmissible through a metallic wire,
he says nothing at all. As, however, his theory, that
the force evolved in muscular action is derived wholly
from electric currents, turns much upon the first of
these points, the question may be deferred till we come
to that subject.

From the foregoing considerations, it seems to me
that the more probable conclusion is that the vis nervosa
is a specific force analogous to electricity, galvanism,
and magnetism, but distinct from them just as these
are distinct from each other, although it is easily convertible into heat or electricity.

CHAPTER VIII.

ON entering on the question of the structure and function of the muscles, let us first give a short summary of Dr. Beale's views:

The transverse markings of the striped muscle, which is the most perfect form of this tissue, have long been objects of attention, but they are not essential to contractility, for they are absent in the involuntary muscular fibres. In the early stage of development of the voluntary muscles, they are also absent, and Dr. Beale states that they do not appear till the act of contraction has occurred repeatedly; and he considers they are due to important changes taking place while the contractile material is in a soft and plastic state. In opposition to the prevalent theory, that muscular movements are the offspring of the protoplasmic movements, Dr. Beale shows that the two kinds of movement are essentially different. According to him " a contractile tissue may be likened to a chain of beads, every bead being capable of becoming short and broad, or long and narrow, but forced to retain its relative position with regard to every other bead. The contractile cord may thus become shorter, causing its points of attachment to approximate " ("Biopl.," p. 214). And he defines a contractile tissue to be one " in which simple movements, like shortening and lengthening, alternate with one another, each movement

9—2

being a mere repetition of the first movement that occurred when the formation of the tissue was complete" (Ibid., p. 237). Such a movement is totally different from the protoplasmic movements described at p. 51, and it is difficult to conceive a substance like the protoplasm contracting as above described. The question of the exact boundaries of the living action, and that of the mere mechanical arrangements for the transformation of force, which together cause muscular movement, is still involved in obscurity. But Dr. Beale has made great strides towards the clearing up this difficulty, and has always advanced in the most philosophical and cautious manner, by keeping the anatomical evidence in the foreground. He demonstrates, first, that for each elementary muscular fibre, whether plain or striped, the sarcolemma, and the sarcous prisms, whether scattered or arranged in Bowman's discs, and also (most probably) the single refracting semi-fluid in which they are imbedded, are non-living, for these reasons :—" The structure of unstriped muscle is smooth, or very slightly fibrous, but exhibits no indications of containing bioplasm in its substance. The tissue is not tinged with the carmine fluid. It possesses all the general characters of formed material, and its relation to the bioplasm is the same as that of the formed material of other tissues. The evidence is therefore against such a view, as regards unstriped muscle. Neither is it probable that in each sarcous particle of striped muscle there is a minute portion of bioplasm, because, in the first place, the living matter cannot be detected at an early period of the development of muscle ; secondly, in inflammation, and in other morbid conditions, in which the masses of bioplasm of tissues are much increased in size, no change is seen in the sarcous particles themselves ; thirdly, the lines of sarcous particles correspond with the wavy bands of the fibrous tissue of tendon, which unquestionably consists of formed material only ; and, lastly, since the very transparent contracting tissues of some of the lower animals do not contain bioplasm in their ultimate fibrillæ, there is good reason for concluding that there is no living matter in the substance of the higher forms of contractile tissue. The phenomenon of contractility characteristic of this class of tissues is therefore probably due to changes in non-living formed material only, and is not in any way dependent for its manifestation upon bioplasm " (" Biopl.," p. 227).

With respect to protoplasm, or living matter belonging to the muscular fibre proper, on the other hand, "the proportion of bioplasm to the formed material in fully-formed muscular tissue is considerably less than in many other textures." "The position of the masses of bioplasm varies very much in different kinds of striped muscle. In some forms we find a row of nearly spherical bioplasts in the very centre of the elementary fasciculus of contractile tissues, in others an oval mass is seen at the side of a very long, narrow fibre, consisting of very few fibrillæ ; and in many of the muscular fibres of various classes of vertebrata, numerous oval masses are situated at short distances, and alternating with one another throughout the whole extent of the tissue within the sarcolemma. This variation in position, and the difference observed in the relative proportion of bioplasm and contractile tissue in muscles which act in the same manner, lead me to infer that the bioplasm is not immediately concerned in muscular contraction" (" Biopl.," p. 225).

In the muscular fibre of the *dytiscus marginalis* a small mass of bioplasm is seen in the centre of each Bowman's disc. The function of these protoplasm masses Dr. Beale holds to be restricted to the formation and repair of the contractile tissue. Besides these masses of living matter belonging to the fibre proper, the sarcolemma upon its outer surface is connected with the delicate intermuscular connective tissue, with capillary vessels, and with nerve fibres. The greater number of the masses of bioplasm on the surface of the sarcolemma of the muscles of vertebrata are those of the numerous nerves and capillary vessels distributed to the elementary fibre ; but they have generally been mistaken for " connective-tissue corpuscles."

So far, then, according to Beale, we have a division into passive or dead fibre apparatus fitted for mechanical work, and living matter solely destined for the formation and repair of that apparatus, but no provision as yet for extrication of force

—a steam-engine without the boiler, as it were.   This is more fully stated as follows :—

"The states of *rest*, of *partial contraction*, and *complete contraction*, are but different degrees of the self-same process of shortening of a delicate fibre.   This contractile fibre, perhaps, consists of a passive basic substance of a fibrous character, through which is diffused a soft material, prone to move in directions at right angles to one another, according to the manner in which external forces operate upon it.   The changing substance upon which the alteration depends can be expressed from the muscular tissue, and coagulates spontaneously, like the fibrin of blood.   Young muscles yield a larger proportion of this material than old ones, but I do not think that it is derived solely from the *bioplasm* of muscle"* (" Biopl.," 212).

* That is to say, in spite of the spontaneous coagulation of this semi-fluid matter, which may be expressed from the contractile fibre, it is still " passive," or dead, and he cannot allow that it can be protoplasm, and evolve force in a vital manner, or be the source of muscular action by protoplasmic movements.   I do not think the circumstance of coagulation after death, although that is one of the signs of living matter, has much weight as a proof of the living nature of the single-refracting matter, for we know the fibrinogen of the blood, and of effused fluids, can remain fluid for long when in contact with living matter, which by its interaction with it prevents the formation of fibrin.   No doubt this is part of the function of the protoplasm-masses (nuclei) of the sarcolemma.   It must be remembered that what Kühne called muscular protoplasm was obtained by shredding frozen frog-muscles freed from blood.   The viscid liquid filtered from this, when thawed, he called protoplasm, but it is obvious it must have contained particles of true protoplasm of the nuclei of nerve, muscle-sheath, capillary and connective tissues.   This would account for any signs of vitality beyond coagulability, which certainly belongs to the muscle-fibre contents, as we see by the rigor mortis.   But lactic acid is also one of the products of death, of probably all true protoplasm—certainly of muscular action—and causes staining with carmine.   Why, then, is the fibre itself not stained ?   It is true that Ranke and Gerlach, from experiments on the muscles of the Axolotl, assert that the intermediate substance between the sarcous elements is coloured by the carmine, while the latter are not.   And Ranke generalizes this for all muscles ; but it is denied by Beale, and this denial is supported by his plates.   I have seen several of the preparations from which these were taken, and I can testify to the accuracy of the plates.   Ranke used preparations made with alcohol, which may have vitiated his results (Henle's " Bericht," 1868, p. 351).

Whence, then, these "external forces?" Here the same philosophical caution is displayed, and no theory entered on till an anatomical basis is discovered. His theory implies that not only the stimulus, but the whole force for muscular action and work is furnished by the motor nerves, and this is founded mainly on the demonstration by his own anatomical preparations of the invariable presence of motor nerve fibres in connection with all muscular forms.

The chief facts concerning the ultimate distribution of the muscular nerves are already given, being included in the general anatomy of nerve distribution. But with respect to the muscles specially, he holds that in addition to the generally accepted plexiform arrangement of the dark-bordered nerve trunks and fibres, and their ramifying distribution to the muscles, numerous far finer fibres proceed from these, and pursue a tortuous course, frequently crossing the muscular bundles. These fine fibres continue to subdivide, and finally terminate in plexuses or loops, never by free ends. Moreover, contrary to anatomists who assert that muscular fibres may receive nerves at one or two points only, and that considerable portion of the ends of muscles are destitute of nervous supply, he asserts—"I have been led to conclude that every muscular fibre is crossed by very delicate nerve fibres frequently, and at short intervals, the intervals varying much in different cases, but I believe never being of greater extent than the intervals between the capillary vessels" ("Biopl.," p. 250). He asserts also, contrary to the opinion of Doyère and Kühne (Stricker, Syd. Soc., i. 207), and others, that the nerve fibres never penetrate the sarcolemma, but always cross and ramify on the outside of it. Likewise, contrary to the opinion of Kühne (Stricker, i. 223), that no general scheme of muscle-nerve distribution can be laid down from its extreme variety in different animals, Beale states—"On the other hand, my observations lead me to the conclusion that the arrangement is in its essential points the same in all classes of animals. In no case are there nerve ends, but always plexuses, or networks, which are never in structural continuity with the contractile tissue of the muscle" ("Biopl.," p. 260).

Finally, that there is no contractile tissue in any animal, from the highest to the lowest, that is not furnished with nerves— *i.e.*, contractile tissue proper—not formless protoplasm or cilia, exhibiting movements. "And as I have detected nerves in every form of contractile tissue that I have examined, I think it right to conclude that contractile textures are invariably associated with nerves" (Beale's "Biopl.," p. 237).

Dr. Beale does not profess to be electrician enough to explain the exact mode in which the contraction is brought about, but is satisfied by showing the existence of a mechanism by which work may be done by induction from a current in closed circuits. "They (the nerves) are certainly not connected in any way either with the nucleus or with the contractile tissue of the muscular fibre. They cross the fibre either obliquely or at right angles ; and oftentimes a nerve fibre runs for some distance parallel with the muscular fibre. The influence, therefore, exerted by the nerve fibre cannot depend upon any continuity of texture between it and the contractile tissue, but is doubtless due to the passage of a current through the nerve, which determines a temporary alteration in the relations to one another of the particles of which the contractile tissue consists" ("Biopl.," p. 239).

This remarkable theory, which is as yet opposed to the great majority of the prevailing opinions, demands all attention, and invites thorough and searching criticism.

It is obvious that the old question of the inherent irritability of the muscular fibre is again raised. Not in the old form, in which the faculty of contracting was supposed to be communicated from the central nervous system, for that has long been set at rest, but whether each individual primitive bundle, of which the muscle as a whole is composed, is idio-muscular or nervo-muscular, in the language of Schiff, adopted by Kühne. Both these authors believe in the contractility of the fibre, independently of nerves; but Schiff thinks the two modes have some essential difference, while Kühne accepts the terms in the sense that the fibre has inherent contractility which may respond to other stimuli than that of the nerves. No one of late, except Fletcher, attributed the faculty to some living matter—not the actual fibre—within the muscle as a whole, till now Beale makes a real and absolute distinction between these two expressions, and denies altogether the idio-muscular actions.

The arguments that irritability resides in the muscle as a whole, independently of the motor nerve trunks and the blood-vessels, are, it is here taken for granted, unanswerable ; but it is also generally assumed that it must reside in the primitive fibre itself, and it is even asserted that the fibre can be seen to contract on the direct application of stimuli, when destitute of nerves. This point demands special examination, and the chief proofs of it rest on Kühne's experiments,* so we had better consider these a little in detail.

The cardinal point is the structure of the sartorius muscle of the frog in respect to the motor nerves. The main trunk of the motor nerve is said to enter about the middle of the muscle, and branch out towards both extremities in ever diminishing size and number of twigs, so that there are five zones of nerve supply, and corresponding irritability. These diminish gradually, and the zones at the extremities possess no nerve fibres at all. Thus two-fifths of the muscle are said to have no nerve fibres, and on this all his results hinge. And, from a variety of experiments, he concludes that "the excitability of the muscle stands in the closest connection with its nerve distribution"

* Untersuchungen uber Bewegungen und Veränderungen der Contraktilen Substanzen. Von Dr. W. Kühne. Reichert and Dubois' "Archiv," 1859, heft. 5 and 6.

(p. 589) ; but that in those two nerveless zones, though the excitability was at a minimum, still with a certain strength of
electric current contraction could be excited, and if the motor
nerves were paralyzed by an upward current, that low degrees
of excitability extended through the whole muscle.   This last
circumstance is difficult to explain on Beale's theory, but
Kühne says it is very difficult to get it to succeed, so perhaps
we may wait till it is confirmed.   In the meantime we may
notice that in Stricker's Handbook, in 1870, Kühne still maintains the assertion of the absence of nerves in portions, and
makes it even more general, describing it as a well-known fact
"that considerable segments of every muscle may be met with
in which no nerves are to be found, and that in particular the
extremities of the muscles appear to be destitute of nerves for
a considerable space" (p. 204).   This, in spite of Beale's well-
known observations—according to which the most important
part of the nerve, as conductor of force, begins where Kühne's
nerve seems to end.

"The active part of the nerve-fibre, as regards the elementary
muscular fibre, commences only at the point where the dark-
bordered character of the nerve fibre ceases, and it therefore
follows that the most important, and most active portion of the
peripheral nerve fibres distributed to muscle, has escaped the
observation of many observers.   The fibres are extremely delicate, and, like other very fine nerve fibres, can only be rendered
visible by special methods of preparation.   Every nerve fibre,
however fine, is compound, being composed of several fine
fibres. 'Nuclei' are invariably found in relation with these
fibres, and they vary in number in different cases" ("Biopl.,"
p. 273).

Perhaps we can account for this discrepancy when we know
that the finest nerve fibres are invisible in the living body, and
Kühne's preparations are directed to be made by macerating
with dilute sulphuric acid, dissolving the connective tissue by
warming to 104° F., isolating the muscular fibres by vigorous
agitation with water in a test-tube, and then pencilling away
the still adhering capillaries with a camel-hair brush.   It would
be difficult to imagine a better way of removing the plexus of

fine fibres described by Beale, and which he rendered visible by
long soaking with all the parts *in situ* in glycerine of graduated
density.* It is plain that if Beale's anatomy of the muscular
nerves is right, the signification of Kühne's experiments must
be entirely changed ; but it must be allowed that there are
some facts difficult to reconcile with the purely passive rôle of
the muscular fibre. Setting aside those resting on irritability
of a supposed nerveless portion, there are the contraction of a
small part when mechanically irritated, which does not spread
to the rest of the muscle supplied by the same nerve. I suppose to this we may say that it is not known how small a portion of the nerve may have independent sources of nerve
power. The fact that if the motor nerve trunks are paralyzed
with curare, no stimulus applied to or through them excites contraction, but if you electrify, or pinch, or bruise the muscle itself, it contracts. This may well be when we consider that
this poison is so specific in character that it has no effect on the
sensiferous nerves ; so it may also fail to act on the intra-muscular nerve protoplasm, which is the source of power ; and, in
fact, Schiff found that irritation of a muscle whose nerve trunk
was paralyzed by curare re-acted beyond the point of contact,
and therefore he concluded that the nerve-ends were not
paralyzed. Again, the action of chemical irritants ; Kühne
(590) adduces glycerine, creosote, lactic acid, and alcohol as
exciting the nerves violently, and, through them, the muscles ;
while, if applied directly to the latter, no action results. If
you apply glycerine to the cross section of the (supposed)
nerveless end, no contraction is excited, while if to the central
cross section, where nerves are, palpable jerkings follow. This
is, no doubt, from the fewness and fineness of the nerve fibres,
and the want of diffusive power of glycerine, for the experiment succeeded only very imperfectly, if at all, with the
creosote, alcohol, and lactic acid, no doubt from their greater
power of diffusion into a region better supplied with nerve

* Tergast, in his memoir on " The Relation of Nerves to Muscles,"
in 1872, denies that any parts quite free from nerves exist (" Jahresbe-
richt," 1873, p. 121).

bioplasts. A more difficult experiment to explain is that of Professor Rutherford,* who dipped the cut end of the sciatic nerve into ammonia without producing muscular contraction ; but when the cut end of the same gastrocnemius was touched with the same solution, immediate contraction ensued : again, when the nerve end, contaminated with ammonia, was cut off, and salt applied to the fresh surface of the nerve, contraction took place. I can only suppose here that the ammonia paralyzed the nerve, while the same quantity applied to the larger moist muscle section was only sufficient to stimulate.

We have also the experiments of Cl. Bernard with Kali sulphocyan : and of Kölliker and Pelikan with upas and veratrin, which seem to show a specific vital action on the fibre itself. The Rhodankalium was tried by Kühne, and, as he states, it only slightly stimulates, and then paralyzes the nerve ; whereas its action on the fibre is very powerful, and when applied to the whole muscle causes violent jerking, speedily followed by death and coagulation (*rigor mortis*). On dipping a portion of the muscle into a solution, that part could be killed, although the nerve passing through it could still conduct stimuli to a sound part. The explanation of these facts is, I presume, simply that the Rhodankalium is a powerful stimulus, soon causing death to the protoplasm of both the fine nerves and the sarcolemma, but that the larger fibres are protected for a time by the medullary sheath.

Unless we can in some way explain the action of these chemical agents, the passive theory of the muscle fibre must be given up, for if they acted at all it must have been as stimuli, and this implies irritability and vital action, for they could not furnish force to cause physical shortening of it.

But a new question is raised by the observation of Wundt,† who states that when the constant current acts directly on the muscle, the contraction lasts and is uniform during the whole time between making and breaking the circuit. And it is thus different from the tetanus produced by electrifying the nerve,

* "Lancet," Jan. 21, 1871.
† Reichert and Dubois' "Archiv," 1859, p. 549.

which is always irregular, and composed of separate longer or shorter contractions. Can it be, therefore, that the galvanic current conducted by the sarcolemma, or connective tissue, may furnish the force inductively to cause the physical change in which contraction consists, just as the nerve force does naturally? We should thus have contraction without vital action or irritability. I cannot find that this observation has been confirmed, and especially if the work done has been measured. Wundt further asserts that the contractile substance itself will not contract under anything else but the galvanic current. Schiff and Eckhard, however, deny that the current can act at all upon it except through the nerves, while Kühne reiterates his assertion of the action of both galvanism and chemical stimuli, in spite of these three. Kühne ultimately, by using very fine electrodes, finds several points in the muscle, even in the centre close under the nerve trunk, where the irritability is at a minimum, just like the extreme ends, and therefore he concludes that these spots are also destitute of nerves. Nevertheless, he observes that these points seem to receive a nerve stimulus in some way, because they contract also when the nerve trunk is excited (594). This harmonizes with Beale's view that they are supplied with the fine nerve fibres which must have been reflected from some finer twigs at a distance from the trunk entrance.

The representation we may keep in mind of the mechanical conditions of muscular fibres in action, is as follows :—To form a fibre (or primitive bundle), we have a long, delicate, flexible closed sheath—the sarcolemma—in which is contained a semi-fluid, single refracting substance, in which are embedded a number of three to five sided prismatic double-refracting bodies, arranged in the striped muscles, in regular layers (Bowman's discs). During the act of contraction these discs become broader and thinner, and approach nearer each other, and, in consequence, the whole fibre becomes shorter and broader, and its extremities are caused to approximate, and thus mechanical pulling work may be done. At the same time, a little of the fluid contained amongst the fibrillæ is forced out, forms vacuoles (Kühne), and lies in bullæ, or blebs, beneath the sarcolemma, which is drawn up into wrinkles (Carpenter). This last circumstance, and the fact that this membrane is always very delicate, and, in some muscles (even the strongest, as the heart) entirely wanting, may let us at once put it out of account in the mechanical agency.

The question is, as Brücke states it, "How can a solid substance, even of the consistence merely of a trembling jelly, really set about the process of contracting itself ?" also, whence is derived the force to do that work ? Kuhne[*] agrees with Bowman, and others, that a wave-like swelling passes along the fibre during contraction when touched, which he thinks shows a state almost of fluidity ; and, in opposition to Kölliker, he says it would be unduly stretching the definition of a solid to call the fibre solid, for it possesses the perfect mobility of the molecules of a fluid, and, in consequence, takes the form which would be given by gravity, therefore it remains after the contracting force is withdrawn pretty much as it would under gravity, and in a state not very different from the contracted shape, unless drawn out by the elasticity or the weight of neighbouring parts. He will not admit the propriety of speaking of a semi-fluid, but says the muscular fibre is a closed tube containing fluid and the muscle prisms. This Beale will not admit, but contends for the state of a soft solid, because some fibres, viz., those of the heart, are destitute of sarcolemma. The action of the flesh-prisms, or sarcous particles, is also very difficult to understand. Brücke (Stricker, Syd. Soc., i.) shows

* Reichert and Dubois' "Archiv," 1859, p. 809.

that they are uniaxial, positively refractile bodies, like rock crystal, and that they are made up of groups of still smaller bodies, to which he restricts the term disdiaklasts. During contraction the flesh prisms, or sarcous elements, become shorter and thicker, and if they were single solid bodies, that would imply a strain and change of position of the molecules, which would alter their refracting power. No such alteration takes place, therefore "the form of the whole group—that is, of the sarcous element—is here changed by an alteration in the arrangement of the several corpuscles, just as in a company of soldiers groups of various breadths and depths are produced by changes in the position of the several individuals" (p. 240).

By the more recent researches of Schäfer, Flögel, Merkel, and Engelmann ("Medical Record," 1873) we find that the shortening of the rod-layer may amount to two-thirds of its length (Flögel). It is denied that the double-refractile rod-like bodies and the single-refracting intermediate substance are distinguishable during life (Engelmann). Each muscular fibre is divided into a number of equal divisions, each of which contains transparent fluid substance, in which is accumulated at the sides of the median membrane the gelatinous contractile substance, the proper transverse stripe (Merkel). "When a muscular fibre contracts, the individual divisions do not pass at once from the state of rest to that of contraction, but they go first through an intermediate state." This latter is characterized by the disappearance of all optical differences, the contents of the muscular division being perfectly homogeneous, and at the same time very bright. "During this state the contractile substance, imbibing all the fluid that is contained in the division, swells so as to fill out this completely, and having gone through this preparatory state, again presses out the fluid and accumulates on both ends of the division beside the terminal disc. The individual particles of the contractile substance press as close as possible to this terminal disc, the former trying to come into contact with the latter by as many of its particles as possible ; in consequence of which the muscular fibre not only becomes broader, but also the fluid which is pressed out by the contractile substance is now accumulated in the middle of the division at the sides of the median membrane. As far as volume is concerned, the contractile substance is in excess of the fluid in a state of rest, whereas in the state of contraction the contrary is the case" (Merkel). To this last sentence Engelmann does not agree, and it is not easy to reconcile the observations of the different experimenters in all details. Probably the above statement of the homogeneity of the division during life refers to a state of complete relaxation. Engelmann also states that during contraction the isotropous substance becomes firmer and less transparent, while the anisotropous becomes clearer and softer. Schäfer, on the other hand, asserts that the ground-substance is anisotropous, while the rods are isotropous and in the state of rest the whole fibre appears anisotropous, because the muscle-rods are surrounded by the anisotropous ground-substance. This last, he assumes, is comparable to a protoplasmic matrix, which is the true contractile part, whereas the rods are elastic structures, which serve to restore the fibre to its original length after contraction has ceased ("Qu. Micro. Journal," April, 1874). So far are observers yet from being agreed as to the physical process.

As to the force requisite to produce these remarkable changes, can we imagine any merely chemical change in the semifluid single-refracting substance capable of evolving and directing it? We are here restricted to the single-refracting substance, which is comparatively small in quantity, for we know that the flesh prisms are not consumed in the process, and how can we imagine an evolution of force from chemical actions in a formless fluid to be so directed as to change so exactly the shape of those bodies imbedded in it? The question seems not worth discussion. The next theory is that these movements are protoplasmic. This, of course, involves the vitality of the fibres, which has already been argued against, and besides that, I think the arguments of Beale are most cogent which rest on the totally different character of the movements, which need not be gone over again. So there seems to remain nothing but the theory of a current force of the nature of electricity, which may be supposed to change the shape of the flesh-prisms by altering the position of the disdiaklasts, or to change the position of the sarcous particles as a whole, and to increase the cohesion of the particles of the fibre as a whole, probably by some change similar to that which iron undergoes when magnetized. This we know becomes lengthened in the direction of magnetization owing, as supposed by De la Rive, to the crystals of which the solid bar is composed setting themselves parallel to the bar in their longest dimensions. Even when the magnetic oxide of iron is reduced to fine powder and suspended in water, in a vessel surrounded with a coil forming part of a voltaic circuit, the same tendency is shown and the particles attach themselves in lines (Grove). If we suppose the action of the nerve current to be the reverse here or diamagnetic as it were, we have a dim representation of the action of the striped and smooth muscular fibres. There is also a clink heard when the circuit is made and broken in the electro-magnet, and this is paralleled by the muscle-sound.

The question, as placed by Dr. Beale, is entirely different from the position which the electric theory has ever occupied before, for now it is not the question of a mere stimulus given by the motor nerves, but the whole force needful for the work must be furnished by them, and as this is a measurable quantitative relation,

10

we must again go back to anatomy to inquire if the organs exist to furnish this amount of force. Singularly enough, Dr. Beale himself seems to have overlooked this point, for I can find no direct allusion to it. We have first to distinguish between the amount of force required for a stimulus and that required for doing the work. The motor nerve trunk, which supplies the muscle as a whole, is certainly sufficient to set agoing the whole apparatus of force extrication within the muscle to its full extent. How much does it contribute to the actual work? The answer is given by an experiment by Matteucci.

After having carefully determined the *minimum* duration of the passage of an electric current through a motor nerve needful to cause the contraction of the muscle to which it is distributed, he measured the force expended in exciting the nerve action by calculation from the weight of zinc dissolved in the voltaic battery during the passage of the current. On the other hand, the amount of mechanical work performed showed the force developed in the muscle when contracting under the influence of the nerve action. On comparing these quantities, it was found that the mechanical work performed by the muscle during its contraction, was equal to 30,000 times the force expended in producing the nerve excitement (Gavarret, " Phenomènes Physiques," p. 255).

To discuss the question further, we must consider the processes necessary in the evolution of the force for muscular work; and for those who may not have the data at hand, I will briefly give the chief facts as yet known :—

It is very interesting to notice how the great truths in science come to bear different interpretations in the progress of knowledge. We are too apt to undervalue the discoveries of our predecessors because the theories founded upon them are subsequently found to be imperfect in some points. Such superficial judgments of one generation are sure to be avenged in the next, for at no time can any department in physiological science be pronounced complete. In this question of evolution of force by muscular action we have had four important principles laid down at different times : 1. " The principle of Liebig, that the oxidation of the albuminous matter of the muscles was the sole source of the force required for mechanical work ;" 2. " The muscle is only the instrument through which the transformation of force is brought about, but it is not the matter by the metamorphosis of which the effect is produced ;" 3. " The enduring power of doing work is proportional, not to the mass of the muscle, but to the mass of blood circulating through it ;" 4. " Not the hundredth part of the oxidative process is performed outside the walls of the blood-vessels." These three last are laid down by J. R. Mayer, in his Mechanik der Wärme (see " Life and Equivalence of Force," p. 50). The last principles appear conclusively to contradict the first one of Liebig, and when we add to that the facts since demonstrated of the non-increase of urea under severe and prolonged muscular work, when the diet was non-nitrogenous, and in addition, the testimony of Beale to the impossibility of the consumption of the actual muscular structure in ordinary work, we perceive that Liebig's principle, in its naked form of statement, has been pronounced untenable with apparent justice. Nevertheless, read by the light of subsequent discoveries, all of the above principles contain fundamental elements of the truth, and therefore constitute important stages in the progress of knowledge. For we come back through Beale's theory to the principle that the muscular fibre is merely a mechanical instrument for the transformation of force into simple work of pulling, while the evolution of the whole force takes place exclusively within what is technically called the muscle, as a whole, by the consumption of a structureless *nitrogenous* matter, which is constantly

10—2

decomposed and immediately renewed from the 'blood circulating within the muscle. This is in harmony with all Mayer's
principles except, perhaps, the last, which cannot be accepted
literally when we consider that in insects the tracheæ extend
into the parenchyma far beyond the corpuscle-holding blood-
vessels, although, if we take it simply to mean that the blood
within the boundary of the muscle is the permanent source of
both the combustible matters and the oxygen, in vertebrates this
will be found compatible with it, taking into account the power
of diffusion of oxygen through the capillary walls, and, what
will be presently adverted to, the storing up of oxygen. Among
the proofs of these propositions we may notice the following :
The idea that force was furnished from the central nervous
system was soon disproved by the observation that muscles will
for a time contract when the connection with it is severed or
even when they are detached from the body. The idea that the
force was produced in the blood and furnished as heat or any
other form of force to the muscle, was also negatived by the
fact that the latter will still contract for a time when quite
empty of blood, although with blood it will do more work, and
that in proportion to the change produced in that blood ; therefore force is not derived from the blood *per se*, but the oxygen
and material with potential energy, whereby the muscle evolves
it, are.

In the resting muscle there is a certain amount of chemical
change continually going on, consisting of processes which
ultimately involve consumption of oxygen and formation of
carbonic acid, but cannot be looked upon as direct oxidation,
but rather a formation by oxidative synthesis of a substance
which splits up into lactic and other acid products which are
neutralized by the surrounding alkaline fluids. These changes
produce the disengagement of active force manifested in the
electric muscle-current and also heat. In the active muscle all
these processes are increased and some other products are recognized, though probably there is no really qualitative difference in
the action (Hermann). More oxygen is taken up and more of
the carbonic and lactic acids are produced, as also of the
alcoholic and ethereal extracts indicating the products of de-

composition of the albuminous and of the crystallizable, saccharine, and fatty elements which occur in muscles. At the same time there is some diminution of the total albuminates, of the watery extracts and of the substances from which lactic, carbonic, and the fatty acids are formed (J. Ranke). It is impossible to go more accurately into the compounds actually found in the living muscle, for of kreatin, grape-sugar, inosite, sarkin, myosin, &c., some are probably, and the last certainly, not formed till after death, and most of the experiments are vitiated by the fact that death takes place during them (L. Hermann). However, in the living state the reaction certainly becomes acid, and there are found products which lessen the capacity of the muscle for work and are called fatigue-stuffs. These are supposed to be chiefly lactic acid, carbonic acid, and phosphate of potash. That the fatigue-stuffs have this effect is shown by the fact that if they are washed out with a weak solution of common salt, and if oxygen be then furnished, the activity of the muscle is restored : likewise, if lactic acid or flesh-extracts be injected into a fresh muscle, the same weakness is produced. Recovery from fatigue therefore depends on the restoration of oxygenated blood and the neutralization of these fatigue-stuffs and their final absorption by the capillaries, the plastids of which decompose them into blood and effete products. During the active state force is evolved and consumed in mechanical work, the electric muscle current diminishes almost to zero and heat is also produced.

The vital and chemical processes during rest and motion are thus not essentially different, but form merely a *plus* and *minus* of the same state. In fact the so-called state of rest is always (in health) a state of minimum of contraction, as is seen by the drawing towards the opposite side, when the nerves of one side are cut or paralyzed. The so-called tonicity is nothing but partial contraction.

Now, although it is certain that the total force for muscular work is derived from the passing down of complex compounds to a simpler state of combination with satisfaction of stronger affinities,* and ultimately may be described and measured as

* " It must not be overlooked that the liberation of force is not de-

oxidation, still direct oxidation is not the immediate means whereby muscular force is evolved, any more than it is of the performance of any truly vital function. "It is also certain that the chemical metamorphoses which furnish the force to the muscles are, in part, not performed at the moment of the muscular contraction" (Ranke, op. cit., 632). This relates chiefly to the supply of oxygen, which is not, as was formerly supposed, derived directly from the oxygen of respiration at the time, but from oxygen already stored up in the organism in some mode not yet settled. The proof of this is that, according to the experiments of Pettenkofer and Voit, the expired carbonic acid is no measure for the quantity of oxygen inspired at the same time, and the quantity of carbonic acid expired during the day greatly exceeds the oxygen inspired during the same time, even if the body be at rest, but still more (to the extent of double) if working, while the contrary is the case during the night. Thus in the waking day the expiration of carbonic acid is due in great part to the surplus of oxygen inspired during the preceding night, and stored up in the system; and in fact the

---

pendent on oxidative processes alone, although this may be the most frequent example of the general law that force is set free by every chemical process through which stronger affinities than before are satisfied. An example of a non-oxidative process whereby, nevertheless, heat is evolved, is given by the alcoholic fermentation of sugar. As shown in the diagram, the affinities of the C atoms, which in the sugar-molecule were unsatisfied either by the carbon or the hydrogen affinities, are after the splitting up satisfied by the O affinities; but as the attraction of O to C is greater than C to C or to H, an evolution of force must take place by this atomic change of position" (Hermann, 191).

$$Sugar = C_6 H_{12} O_6$$

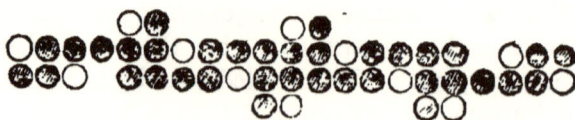

Splits up into carbonic acid and alcohol

$C_2 H_6 O$ CO_2 CO_2 $C_2 H_6 O$

organism is more vigorous and fit for work the more oxygen is thus stored up.

It was also observed by Valentin, that hybernating animals at times gain weight between the weighings, owing to absorption of oxygen, although a certain constant loss by carbonic acid and water is going on. Also in diabetes and leucœmia, in which the debility is so well known, Pettenkofer and Voit found that the nocturnal absorption of oxygen does not take place. Thus is accounted for the fact that labourers are unable to do the same amount of work on a diet with too little flesh-meat, although furnished with an ample supply of the fatty and starchy matters to extricate the requisite amount of force. In this case, as in disease and in otherwise unwholesome feeding, the requisite nocturnal storing up of oxygen does not take place to render the individual fresh and vigorous in the morning.

Pettenkofer and Voit showed that this store was formed only through the means of albumin-substances, and the maximum of the storing was regulated by the amount of these. So even if the ultimate oxidation of fats and sugars did furnish force for work, still the albuminates by this storing of oxygen always took an essential part in the process, and therefore "in this case also the quantity of the albuminous matter determines the amount of work"* (p. 14). Again, after disproving Liebig's 2nd and 3rd propositions, viz., that muscular work is the sole cause of decomposition of albumen, and urea the measure of work, and that it is the formed muscular structure which is consumed, Pettenkofer and Voit still uphold the 1st proposition, viz., that *the work of the muscle is furnished by the decomposition of albuminous matter alone.* This, in spite of the difficulty that the urea is not increased by muscular work ; and Voit controverts by experiment and reasoning all the efforts to explain this away, even those of Liebig who suggests that the products are not excreted in the same day, and appeals to the experiments of Parkes, who, however, drew no such conclusion from them, but, on the contrary, confirms the fact that there is no increase of urea, though he gives no help in the ex-

* "Lehre von der Quelle der Muskelkraft," &c., von Carl Voit, 1870.

planation of the fact. Voit, in effect, says that work is done as in many machines, by a store of force laid up at some other time, as by winding up a spring for example, this store being laid up by the intermediation of albumen binding the oxygen and then splitting up ; but he does not make it clear how this takes place without increase of the urea according to the work, for the more work a machine does the more winding up it needs, no matter when.   Voit, besides, shows that fat and milk are formed from albuminous matters and not from the non-nitrogenous principles, and he rather falls back into the old ther-mogene theory, for he says : " In a pure albuminous diet the fat produced from it serves for heat production, while by introduc-tion of easily combustible carbo-hydrates into the diet, these last are oxidated and the fat originating from the albumen remains accumulated " (p. 81). Voit resembles Goodsir and Beale in describing " milk as a dissolved organ of the body, and not a simple filtration from the blood " (p. 83), and that in the secretion the mammary gland uses albuminous material for the building up of cells, which then partly undergo fatty degenera-tion and partly take up fat from the blood.   In respect to the difficulty of the non-increase of urea in proportion to work, the explanation of Kühne is probably the one with which we must be satisfied for the present, viz., that compensatory pro-cesses are taking place at the same time, of the exact nature and amount of which we are at present ignorant, but which may " decompose an absolutely great amount of albumen while pro-portionately little is oxidated down to urea, and therefore the excretion of the latter is as slightly increased during work as experiment shows."*

A single yeast cell will convert sixty times its weight of sugar into carbonic acid and the still combustible alcohol with evolution of active force as heat, not by catalysis, for there are other products, and the cell either grows or wastes away accord-ing as it is furnished with a due amount of nitrogenous pabu-lum. In this case some compensatory processes must have been at work, and the same particles of nitrogen must have taken part in the vital decomposition and recomposition over

* " Physiologische Chemie," p. 327.

and over again. A bee will move with extreme activity upon a diet of pure honey or sugar, and thus by a more complex organization convert the chemical force of carbo-hydrates, by union with oxygen, into motion, which the yeast cell could not do, while its tissues do not waste, though supplied with so little albumen, thus showing the existence of compensatory processes. In spite of the supply of oxygen to the extreme points by extension of the tracheæ beyond the blood-vessels in insects, no mere oxidation can be the cause of evolution of muscle-force, for the theory that an animal could convert the heat of carbo-hydrates into motion, though still taught in some text-books, was long since disproved by Clausius and Voit, followed by Fick. As to where this storing up of oxygen takes place, no doubt it may be partly in the blood, which must be richer in circulating albuminous matter, and no doubt also part of this store is consumed in oxidating and removing effete products, but the chief part must be in the muscles themselves, for a muscle when detached from the body, with all the blood washed out, and when placed in a vacuum or in indifferent gases, will still contract and do work for a time. And in the living body, J. Ranke states, " I could prove that the capacity for doing work of the muscle rises and falls with the amount of solids contained in it, so that a muscle is the more capable of work the richer it has been in normal muscle-stuffs during the state of rest" (635). Likewise, it is known that the amount of carbonic acid in the venous blood of the active muscle is greater than can be accounted for by the disappearance of oxygen.*

During work from stimulation by voluntary or reflex motor nerves, both the formation of this consumable matter by oxidative synthesis from the blood and its dissolved oxygen, and the consumption of it are immediately increased. But this increase does not equal the loss, and consequently the store wastes, while the neutralization and removal of the acid products does not keep pace with their formation, and therefore the fatigue-stuffs accumulate—hence fatigue, and recovery after

---

* The venous blood of the active muscle contains 3 vol. per cent. less oxygen and 4.1 per cent. more carbonic acid than that of the resting muscle.—Sczelkow.

the needful rest.   J. Ranke is inclined to give the whole causation of fatigue to the accumulation of fatigue-stuffs, and little or none to the over consumption and modification of the protoplasm.   But I cannot agree to this, as fatigue, after over-stimulation, is universal in all living functions, and especially those of the nervous system, while we have few proofs of accumulation of fatigue-stuffs.   Besides, we know that when one kind of stimulus has ceased to act, another will do so—certainly not from removing fatigue-stuffs.   This harmonizes with the theory that storing up is partly a simple growth of the protoplasm whose decomposition is the source of the nerve force which does muscular work, although there is a possible intermediate chemical stage here ; still in all vital action the same vigour and freshness is felt after repose.

It is certainly noteworthy that a veteran experimental physiologist like Voit should have the conclusion forced upon him, that muscle-work force, fat, and milk are all, even in herbivorous animals, produced by the decomposition of nitrogenous matter, and it is so far in harmony with the hypothesis of Fletcher,*

* Dr. Fletcher (" Rudiments," &c., p. 108), after setting aside all the theories of active elongation, reviews under six heads all the theories of muscular contraction up to that time (1836), and concludes that the most probable physical cause was the zigzag bending of the fibre, though he rejects the hypothesis then tacked to it, that these bendings occurred where a nerve loop passed round the fibre, and that the drawing was caused by galvanic force sent by the nerve trunk.   It is curious how nearly Beale comes back to this, but at that time, the electro motor force of galvanism was hardly known, and its conditions were not practically understood.   Fletcher did not attempt to explain how the vital was connected with the physical action.   He says, " But, at the same time, it must be continually kept in mind that there is still a wide, unoccupied gap between simple irritation or the perception of a stimulus by a muscle, and the assumption by its fibres of this zigzag direction as a consequence of it."   Nevertheless, this irritability is not inherent in the fibre itself, but in the ganglionic nervous matter [protoplasm], interwoven with it, for in accounting for the essentiality of a supply of blood to contraction, he says, " we must remember that it is by the blood that the nutrition of the limb is maintained, that its ganglionic nervous tissue ceases in this case [tying the arteries] to be renewed in proportion as it is exhausted, and that there is consequently, after a time, a cessation of irritability—a necessary condition of muscular contraction " (117).

that every manifestation of vital action is due to the consumption of "irritable matter" — a nitrogenous substance in a peculiar state of combination—and also with the identical hypothesis of Beale in respect to "germinal matter," both now corresponding to protoplasm. It is quite possible, however, that the actual evolution of force may take place in an intermediate chemical compound formed by the protoplasm, and that may be even of such a nature, according to the hypothesis of Traube and L. Hermann, as to allow the compensatory process to take place within the muscle, and that this explains the non-increase of urea. However this may be, the difficulty here is no greater than for the whole doctrine of Beale, viz., that all nutrition and secretion are produced by the death or decomposition of protoplasm—*i.e.*, a nitrogenous substance—while many of the products are simple binary or ternary compounds. We are obliged to assume that in the perfect and complex machinery of the living organism, the molecules are arranged and re-arranged through a long series of still unknown compounds, so that the more simple and effete compounds only are finally excreted and thus the nitrogen is retained. The facts are also in accordance with this, for with a diet of purely nitrogenous matters, in many animals all functions can be performed, and heat and work kept up, and the formation of fat and other secretions goes on. In such a diet, then, is the maximum of urea excreted; but, as we substitute non-nitrogenous articles, these are consumed and their products excreted, and the quantity of urea comes down to a certain minimum, below which it cannot fall, however rigidly non-nitrogenous the diet be. Independently of this, the excretion of urea is almost entirely regulated in health by the quantity of nitrogenous matter in the diet, no matter the amount of muscular work done. Thus, the immediate source of muscular, as of all other work and every other vital act, may be the decomposition of a nitrogenous compound, such as protoplasm, yet from the harmony and perfection of the working of the organism as a whole, the most perfect economy in extracting the force of the food is attained by the excretion of only the spent products of the diet, whatever its nature.

The question now is, where do all these processes take place in the muscle ? The parenchyma is now pretty well explored, and there are no considerable territories still unknown in which these operations may take place. According to Beale, the change of matter must take place in the protoplasm, but the bulk of the muscles, *i.e.*, the fibres, do not consist of protoplasm and are not living, nor is the power evolved in the bioplasts of the muscular fibre. There remains only, therefore, the protoplasm of the motor nerves or of the capillaries or of the connective tissue. The protoplasm of the capillaries (venous at least) is no doubt fully occupied in the recomposition and absorption into the blood of the products of change, and the connective tissue corpuscles are too insignificant to be taken into account. Therefore, we are compelled to fall back upon the protoplasm of the intra-muscular motor nerves (and, possibly, to some extent, the bioplasts of the arterial capillaries) as the *source of the whole power of muscular motion*, and unless there is an anatomical basis corresponding to this function, the theory cannot be upheld.

As before said, Beale has not adverted to the quantitative relations of nerve action in muscular work, but when we look into his anatomy we find even a stronger testimony to the correctness of his doctrine, in that it is in a manner involuntary. This lies in the size and number of the nerves connected with the muscular fibres, and of the protoplasm masses belonging to them, and incidentally this comes out in the discussion of the endings in tufts and nerve plates given by Kühne. If we consider these all to be masses of nerve-protoplasm and, as such, sources of electric or other force evolved by nerves, we can see a sufficient anatomical basis for the actual work of the muscles being performed by the motor nerves within the muscles.

It is hardly necessary to add to what has been said on the general abundance of protoplasm in the peripheral parts of the nerves, and especially of the motor nerves in the muscle by Dr. Beale, figured in his plates and quoted at p. 134 ;* but we proceed at once to notice the question of the alleged nerve "tufts" and "eminences."

These have been brought under the notice of anatomists by Kühne specially, who still adheres substantially to his original opinions in Stricker's "Handbook" in 1870, notwithstanding the published discoveries of Beale in 1864, 1865, and 1868. Again Dr. Beale reviews the whole controversy in 1872, and as nothing of moment has been added since, it is unnecessary to

---

* In addition, I may call attention to the size of the ganglions in the auricle of the Hyla and of the local reflex ganglions described and figured by Beale, and also similar ganglions belonging to the capillaries of the frog's bladder described and figured by Dr. Darwin, in the " Qu. Micro. Journal," April, 1874. It would seem that these were intended to furnish the whole force of muscular contraction, and not merely the infinitesimal quantity required for stimulus.

go into the question in detail. It is sufficient to say that, according to Kühne, in many of the invertebrata and the amphibia the motor nerve ends in a sort of tuft or eminence upon the sarcolemma at one or more points, and that here the sheath of Schwann becomes continuous with the sarcolemma, so that the axis cylinder thus penetrates the sheath and comes into contact with the fibre itself. Here it frequently spreads out into a branched end plate, or "layer of protoplasmic muscle substance that may stretch to a variable extent into the contractile part of the fibre" (209). Rouget also describes in lizards a mass of nuclei and granular substance, beneath the point of entrance of the nerve. Kühne states that the nerve eminences vary much in form and size in the Reptilia. In the Lacerta agilis the plate he gives of the "terminal nerve plate," or the "motor nerve plate," is copied into many text-books, and the nerve is here seen terminating in a large, non-granular, thin layer. The thickness of this is considerable, and in the central part is nearly as great as the short diameter of a nucleus of the basis substance (223). Kühne concludes, that although there is so much variety in the termination of motor nerves in muscles, that no single scheme can comprehend them all [this is denied by Beale, as said above], yet, "The extremity of the axis cylinder always corresponds to a remarkably broad expansion, which constantly forms a flat branching mass" (227). Beale, as we have seen, denies that the axis cylinder ever penetrates the sarcolemma, but admits the existence of structures resembling the above outside the sarcolemma; he says, also, that although the latter is drawn into a sort of eminence at the point where the nerve begins to ramify over the fibre, that eminence is not a special organ. With respect to the "end plates" and "tufts," Beale has not found them in the muscles of animals generally—in fact they are rather exceptional—though he has found them in the lizard and chameleon, and that they are not terminal organs, but display fibres passing both to and from them. In fact these tufts are reduplications, and expansions, and coils of nerve-fibres interspersed with large masses of bioplasm, which last are conspicuous in Beale's plates, xii. and xiii. ("Biopl.," p. 263). Beale's conclusion respecting the "nerve tufts" is as follows :

"'Nerve tufts' *are not terminal organs but networks.* The nerve tuft consists of a complex network of fibres, the meshes of which are very small. Connected with the fine nerve fibres are numerous masses of bioplasm or nuclei. The plexus or network constituting the nerve tuft is not terminal, nor does it result from the branching of a single fibre, as has been represented. *Many fibres* enter into its formation ; and from various parts of it long, fine fibres pass off to be distributed upon the surface of the sarcolemma. It seems most probable that at the situation of these compressed coils (nerve tufts) the contraction of the muscular fibre would commence, and that from the nerve current traversing several fibres collected over a comparatively small portion of muscle, the contraction at these spots would be sudden and violent, while it is probable that the contractions commencing at these points would extend, as it were, from them along the fibre in opposite directions. I consider these nerve tufts, therefore, simply as collections of nerve fibres, differing only from the ordinary arrangement before described, somewhat in the same manner as the compressed nerve network in a highly sensitive papilla differs from the lax expanded nerve-network in the almost insensitive connective tissue" ("Biopl.," p. 268).

These organs were found numerous in the tongue of the chameleon, and in corroboration of the supposed use of them, I may notice the function of that organ :

"The chameleon is another curious example of a reptile obliged to employ its tongue in securing insect prey. The chameleon is arboreal in its habits ; its feet, cleft as it were into two portions, firmly grasp the boughs upon which it climbs ; while its well-known power of changing the colour of its skin, so as to imitate that of the branches around it, efficiently conceals it from observation. The tongue of this creature, when extended, is as long as its whole body, and is terminated by a club-shaped extremity smeared over with a viscid secretion ; when an insect comes within a distance of five or six inches from the chameleon, the end of this tongue is first slowly protruded to the distance of about an inch, and then, with the rapidity of lightning, launched out with unerring aim : the fly, glued to its extremity, is with equal velocity conveyed into the mouth" ("The General Structure of the Animal Kingdom," Rhymer Jones, p. 687).

Thus we have in the anatomy of Dr. Beale a sufficiency of source of power, or nerve-force, for muscular work in these intra-muscular masses of protoplasm, so profusely distributed in connection with the equally abundant supply of nerve fibres, at least as far as can be judged by inspection alone. And, as yet, no means are known of testing such a question by actual experiment. At first sight, as the bulk of the muscle consists of the contractile fibre, one would think it unlikely that the source of the power should be elsewhere, but when we reflect that the flesh prisms which form the bulk of the fibre cannot be the source of the power by which they are moved, we have only to look to the single-refracting matter, and even if the consumption of that were a possible source of regular movements, its quantity must be smaller than that of the nerve protoplasm.

Whether the evolution of force takes place by direct metabolic or vital changes in the living matter itself or in a chemical substance of high potential energy produced by it and decomposed as required, presses itself on our attention here. In commenting on Hermann's chemical theory in my former work, in 1871, I suggested that this might be the case, and I perceive since then that Dr. Beale has expressed the same idea, for he says, at p. 205 of "Bioplasm," that the nerve current originates in the bioplasm, or in the "*soft formed material on its surface.*" And again, "by *chemical* changes in the *matter formed* by the bioplasts electrical currents may be produced and then traverse the fibres." While the function of the formation of this substance and the means of decomposing it when required, would still remain with the true living matter, there is room here for subordinate, merely chemical action in aid of the functions of the living body as a whole, such as we see also in the action of the digestive juices. It is possible, also, here occurs a portion of the storing up of force-producing material, and that here also, and not within the actual fibre, we have scope for some hypotheses such as those of Traube and Hermann, which, however, need not be gone into, especially as the last involves the incipient formation of myosin, a proximate principle, which certainly does not appear till after death, and is the cause of *rigor mortis.* It is possible, also, that some share in the formation of this chemical substance may be taken by the protoplasm masses of the arterial capillaries.

Nature of the Nerve Force, and its Relation to Electricity.— Since the time of Galvani, the existence of the nerve and muscle electric currents has been known. It has since been ascertained with accuracy that in the living muscle and nerve, during rest, a current is detected when connection is made between the cross section and the longitudinal surface of the muscle or nerve: the electricity of the former being negative, and that of the latter positive. Also two points at unequal distances from the centre of the longitudinal surface are positive and negative towards each other, and a different position of these poles is taken up when the section is oblique. Several other minute points respecting the behaviour of animal electricity are also known, but as their signification is as yet quite undetermined it is unnecessary to particularize them. As the muscle is surrounded with moist conductors, these electric currents are continually brought into equilibrium throughout the

11

body and converted into heat during rest. When the muscle (or nerve) acts the current immediately ceases, or undergoes the *negative variation.* This is not, as C. B. Radcliffe explains, a change to a current of negative electricity, but simply a weakening or exhaustion of the strength of the current almost, but not quite, down to zero. This negative variation is so constant that, in default of the visible sign of action which the muscle gives by contraction, it is taken as a sign in the nerve that it is in the acting state. "Nothing is as yet certainly ascertained respecting the quantitative relation of the negative variation to heat production and mechanical work. The variation takes place in the stage of latent irritation (of Helmholtz), and only occupies a very short time, *i.e.,* 0.001 of a second" (Bezold). Hermann, p. 239. The stage of latent irritation is the period which elapses between the application of the stimulus and the contraction of the muscle. A certain time thus appears to be required for the change from the production of the electric current to that of work. The muscle current belongs exclusively to the living muscle, and gradually diminishes after somatic death till it quite ceases when *rigor mortis* comes on. It is also stronger the more functionally capable the muscle is, and is diminished by privations, poisons, and all agents which injure or destroy life. The same applies to the nerve current.

These are the chief facts of importance in respect to animal electricity, and they form a slender foundation for the theory of the identity of that force with the nerve force. The disappearance of the current on the action of the muscle does not necessarily show that the motor force of the muscle was electricity, for the same would happen, whatever the nature of the nerve force, if it were directly transformed into mechanical work instead of into electricity as it is in the resting muscle.* The theory, in fact, rests more on analogical

* The negative variation on the occurrence of contraction would at first sight appear to speak in favour of Dr. C. B. Radcliffe's ingenious

reasoning, as we have seen, p. 127, and also on the action of electricity itself on the nerves and muscles. This last is a wide subject, and one for which a volume would scarcely suffice; so I will not touch upon it further than to notice that, besides the difficulties of the electric actions—which are merely physical, and would take place in any moist, imperfect conductors— we have its action as a stimulus, and the whole forms a complexity so great that little practical knowledge has as yet been gained.

One thing, however, must be noticed, viz., that by Beale's theory the question is put on a different footing, for the muscular fibre is put out of account as the recipient of a stimulus, and the nerve protoplasm alone has to be considered directly, and when we speak of irritability it is only of the nerve protoplasm.

The action of electricity is thus to stimulate, as any

---

theory ("Nature," Jany., 1872). He holds that muscular contraction, whether in ordinary muscular contraction or in *rigor mortis*, is nothing more than the result of the operation of the elasticity of the muscle upon the discharge of the charge of static electricity which had previously kept up the state of relaxation. That the two opposite charges are disposed Leyden-jarwise upon the two surfaces of the sheath, and cause elongation of the fibre by compressing between them the elastic sheath, and the contraction of the fibre follows the discharge of these charges by the operation of the elasticity thus brought into play. This theory is so totally opposed to what is here held respecting the nature of *rigor mortis*, and of the sarcolemma, and the fibre, &c., that it is unnecessary to go into it at length. It has recently been found by Dr. B. Sanderson that there is an electrical current from the proximal to the distal end ꞏof the living leaf of the *Dionæa muscipula*. If a fly creeps in to the sensitive hairs and causes the leaf to close on it, the needle swings to the right, showing the negative variation. The author adds, "But it is to be borne in mind, that although when the muscle or leaf contracts, electro-motive force disappears and work is done, there is no reason for supposing that there is any conversion of the one into the other, or that the force exercised by the organ is electrical' ("Nature," June 18, 1874).

11—2

other stimulus, the nerve-protoplasm, and through it effect muscular contraction. This it certainly does, and it also produces all other effects which the moderate and excessive action of stimuli do on living matter: it also produces a variety of electrical phenomena which are not connected directly with life at all. But I can find no experimental evidence, except that of Wundt (see p. 139), which seems to show that electricity (from without) can cause the muscular fibres to contract independently of the nerves, and no quantitative experiment at all showing that it can furnish, to any measurable amount, the force for muscular work. Under greater stimulus, we know, more work can be done; but the relation between that increase is not known, while the relation of the stimulus as a whole to the action performed is infinitesimally small.

Dr. V. Poore, in the "Practitioner" for Jan., 1873, has called attention to the power of the galvanic current in refreshing the muscles and removing the sensation of fatigue, as had been noticed already by Haidenhain and Cyon. Dr. Poore found that when a man was made to hold out a weight at arm's length, the feeling of fatigue which quickly supervened was at once removed on passing a mild galvanic current through the nerves of the arm, and that, whereas he could only support the weight for six minutes when no electricity was used, he was able to support it for more than double the time when the muscles were refreshed now and again with the galvanic current. Further, on testing the strength of the muscular contractions, it was found to be considerably greater when the galvanic current was being used. But no experimental evidence is given as to whether any part of the "mild galvanic current" accounted for the work done, and its action was most likely that of an additional stimulus both to nerve action and the removal of fatigue stuffs. Therefore, in the meantime, we may take electricity into account solely in its capacity of a stimulus.

When a galvanic stream is sent along a nerve the latter is thrown into the state called "electrotonus," which, however, is a merely physical state, and may be represented by any moist cord except for the changes caused on the protoplasm of the living nerve by the action as a stimulus. Thus, in the state of electrotonus, if moderate, the irritability may be increased just as under moderate heat, while by excessive electric action it is paralyzed, or destroyed altogether. The most powerful action as a stimulus is produced by changes of intensity of the current. Besides these actions as a vital stimulus, it appears that as a physical force an opposing current may neutralize and stop the nerve current (which, whatever its nature, is also probably a physical force), but whether physically or as a paralyzing agent seems not determined.* In addition to these there are an immense number of facts known which have probably little

* From the recent experiments of Mr. Dew Smith, of Cambridge, the influence would appear to be physical ("Journal of Anat. and Phys.," Nov., 1873). Mr. S. stimulated the sciatic of a frog by electricity simultaneously at two points, one farther from the muscle than the other. The force, he observes, of the muscular "*contraction resulting is the same as when the near point only is stimulated:* but when a small interval of time is allowed to elapse between the two stimulations the contraction increases, rising to a maximum as the interval is lengthened, and afterwards dividing into two independent curves." This is not merely a variation of the phenomena of summation of contractions as observed by Helmholtz, for these apply to maximum shocks only. The experimenter thinks it implies "a block of nervous impulses," for a current proceeds in both directions from the two points stimulated, and the two impulses moving in opposite directions mutually antagonize each other. This, I think, points more to the neutralization of movements of a physical force rather than to a paralytic vital influence, and it may throw light on the nature of inhibitory nerve action. —

bearing on the physiological question when we recollect the
infinite variety of merely physical phenomena which may be
displayed in such a medium as the animal body when subject
to electric currents, considering the many new facts discovered
in making telegraph cables even.

We come back to the question by what force, developed in nerve protoplasm and conducted by nerve
fibre, the whole of muscular work can be done as supposed by Beale. If some of the foregoing facts favour
the idea that the force is electricity, yet we must remember that the nerves are not better conductors of
electricity than other tissues. "The moist tissues,
with the exception of the bones, are all equally good,
or, rather, equally bad conductors of electricity—about
three million times worse than quicksilver" (J. Ranke).
Besides, as before said, there is no demonstrable insulation by the sheath of Schwann, so the nerves are not
fitted for simple conduction of electric currents; and
these have no reason to choose the nerves as their
channels, so they spread through the moist tissues
almost uniformly.* These facts form an insurmount-

---

* If the two electrodes of any current apparatus are laid on any
two given points of the human body, the electricity from the positive
pole spreads through the body in all directions, to be gathered together
again at the negative pole. The whole body must be looked upon
theoretically as filled with current curves. Hence it appears as if it
were impossible to submit any isolated spot to the electric influence.
In fact, strictly speaking, it is impossible, and the fact cannot be
sharply enough insisted upon." Practically, the electrization of isolated parts is only effected in a very qualified degree by the greater
strength of the current at the spot over that which spreads by diffusion (Fick, "Die Medicinische Physik," p. 370). This, no doubt, besides interfering with electro-therapeutics, vitiates many physiological
experiments. Dupuy found that in electrifying parts of the brain the
stimulus was widely diffused over the hemispheres, and the results did
not correspond with those of Dr. Ferrier ("Lancet," 24th January,
1874.

able argument against Dr. Beale's hypothesis that electricity is the nerve force, and exactly his theory of muscular action adds insuperable force to the argument against the electric theory, for, although it might not have mattered so much sending the force of a mere stimulus, of which so small a quantity is needed, through a bad conductor, how can we reconcile it with the economy of nature, that a force which is to do the whole work of the muscles should be sent through a conductor which offers three million times the resistance of mercury, and still more than that of silver or copper? We are, I think, compelled to conclude that the force must be a distinct force, not like heat, light, or sound, but a current force, analogous to electricity, galvanism, and magnetism, but distinct from these. Perhaps this is what Dr. Beale means, and indeed physicists generally describe these all as modifications of electricity. Nevertheless, Professor Challis regards these forces as essentially distinct. He says:

"It is, I think, to be regretted that experimentalists use the word 'electricity' with so great a latitude as to its application. The practice seems to have arisen from speculatively inferring the identity of different physical forces from phenomena which they have in common, and not giving sufficient consideration to other circumstances by which they are distinguished." He has "uniformly regarded 'electricity' as an effect produced by friction or other means exclusively at the *surfaces* of substances," and he considers it quite distinct from galvanism. He adds, "I have with satisfaction noticed that the Astronomer Royal employs in the Greenwich observations the terms 'galvanism' and 'galvanic signals;' but *electric* telegraph is, I fear, too established by common usage to be unsettled," p. 99. He looks upon the galvanic current as very different from the electric, but nearly resembling the magnetic, into which it is easily converted. He rejects the term electro-magnetic, and would substitute 'galvano-magnetic." The currents which act upon the galvanometer are produced in three ways, viz., by chemical action, by magnetic current, and by heat. These he would designate chemico-galvanic, magneto-galvanic, and thermo-galvanic ("Mathematical Principles of Physics," 1873).

Therefore, I think it would be better to speak of the influence in question as nerve force, or *vis nervosa*, or neuric force, and adopt the terms neurolysis and neuric induction to express the functions corresponding to electrolysis and electric induction, if such exist. I do not think psychic force is a happy term, for that implies something mental, while mental action only expresses itself through this very nerve force, which can exist quite well without mind.

Now, if we admit the theory of Beale, that the muscular fibre is passive, and the whole work of animals is done by a current force analogous to electricity acting through that passive fibre, we come to a somewhat remarkable conclusion. When we consider the known modes in which work is done by the forces generally called electric, we perceive that all mechanical work is done by induced action. The telegraph, and all electro-motive machines, are so worked. While no direct work is done by galvanism, except chemical decomposition. Supposing the nerve force to be similar, this would be represented in the living body by the statement: All action of the nerve force as a stimulus within the nerve circuits, corresponding to electrolysis, or neurolysis, as it might be called, involves a vital change, while all *mechanical work* of the nerve force is by induction of a merely physical change on the muscle fibres. This harmonizes marvellously with Beale's theory of muscular action, whereby fine nerves in closed circuits charged with neuric force cross the fibres in contact with them but outside the sarcolemma, instead of plunging into living actively contractile matter and ending there as Kühne and the majority maintain.

But with respect to the action of the *vis nervosa* as a stimulus, according to Beale, we must admit two distinct modes, viz.; 1st, that of the afferent and efferent nerves in which a continuous closed circuit between the peripheral and the central protoplasm is formed, and the nerve cords act by contact with the living matter: in fact in what would correspond to electrolysis in chemical action, or neurolysis as we may say; and 2nd, by some induced or derived influence if a stimulus is directly given to any other living matter outside of the closed nerve circuit. This last mode also seems to be now admitted by Dr. Beale.* We have thus not the same simplicity as in the more common

* In Dr. Beale's memoir on the relation of nerves and pigment, and other cells ["M. Microsc. J." Feb. 1872], he re-asserts that whatever be the nature of the influence produced by the nerves upon the structure and the action of various tissues and organs, it is not dependent on continuity of substance between the nerve and the tissue affected. But he adds now that when *contraction of bioplasm* follows upon irritation of nerve fibres the result is due to a change in the nerve which runs *near* to it, and not to an influence propagated to the bioplasm from the nerve by reason of continuity of material. Though he here denies that the nerve fibres are continuous with the connective tissue corpuscles of the cornea, and with the pigment cells of fishes as maintained respectively by Kühne and Pouchet, yet this is an admission of the vital action of nerve force on protoplasm as a stimulus and is an abandonment of his previous position, and allows the whole doctrine of trophic and secreting nerve influence. For be it observed, it is here a vital act under nerve stimulation that is in question—not the physical contraction of a muscular fibre which is always pronounced to be dead.

theory that all nerves pass from living nerve centres as single cords, and terminate or begin in contact with living matter, whether glandular or tissue, including muscular and sense organs. There is, however, also a simplicity in Beale's view, if we represent it thus : Nerve cords are never in continuous contact with anything except nerve protoplasm, central or peripheral, and all the work done by the nerve force out of the circuit of the nervous system, is inductive, although that may be employed in two ways, viz., as a physical action in shortening the muscle fibre, or as a stimulus to the living matter of tissues and organs. But the question is not one of simplicity or symmetry, but one of truth and fact. And if Beale's mode is not incompatible with trophic and secreting nerve influences, we must recollect that Pflüger's mode is not the only possible mode of communicating nerve influences, so neither can be decided upon by *à priori* reasoning, but solely by experiment and observation. The question, as an anatomical one, cannot be said yet to be settled, although I think Dr. Beale has, as yet, successfully met all the objections to his scheme of nerve distribution.*

* In spite of Dr. Beale's writings and preparations the termination of nerves in single cords is still generally held by anatomists. One of the latest writers, Dr. Thin ("Journal of Anat. and Physiol.," Nov., 1873), describes the tactile corpuscles as representing the termination, each of a single medullated nerve fibre. And, as before said, the termination of single nerve cords in glandular, muscular, and other protoplasm is still the prevailing anatomical doctrine. Without questioning the accuracy of Dr. Beale's anatomy, that in all cases more than one fibre will be found connected with every nerve bioplast, and that loops are the only mode of distributing nerve force to other parts than the nerve bioplasts, central or peripheral, which are in continuous contact with the fibre, yet I think he is hardly explicit enough in his theoretical view of the function of the second cord, for he does not

In conclusion, I have gone into this lengthened commentary on the nerve and muscle theory from respect to the high authority from which it comes, and its intrinsic merits, and as it is held to be a test of the theory of the sole vitality of the protoplasm.   For it is the most vulnerable point of Dr. Beale's views, and far from being yet sufficiently proved.   But if ultimately disproved, I do not think the protoplasmic theory would thereby be affected, for vitality would not necessarily be attributed to anything rigid and possessing structure, but only to the isotropous semi-fluid contents of the muscular fibre sheath ; we would simply have to accept the muscular fibre as containing protoplasm, and the contraction to be a vital act just as secretion is.   In other respects the protoplasmic theory would remain the same.   The common theory has at first sight nearly everything in its favour, for all protoplasm is contractile, and gathers itself into a ball under stimuli ; the muscular fibre also contains what to the microscope appears to be protoplasm, and is said even to take the carmine (see p. 132).   And it is

---

state whether it is a return cord essential to complete the circuit, or whether it or they, if more, are commissural or afferent to different nerve centres.   The whole subject of electro-dynamic actions of currents not closed, but with ends as in the Leyden-jar, is not yet worked out by the mathematical physicists to the degree of accuracy that has been attained with respect to the closed galvanic circuits, so perhaps we can hardly expect any decisive theory of nerve action as yet.   But Dr. Beale appeals to the action of the electrical eel, the force of which is certainly electricity, as an example of the mode of transmission of nerve force.   It is not very clear how the shock of this animal is transmitted, but it does not appear that a circuit is formed through the animal of which the recipient forms part.   Again, in the muscular nerve circuits he does not state where the return cord of the nerve loop is inserted ; whether all return to the spinal cord, or whether some return to protoplasm masses within the muscle itself.

only after many considerations, such as are given above, that the strength of these arguments is weakened, and the power of a substance so constituted physically to exert such force is discredited.* On the other hand, examples are wanting of a dead substance physically so constituted as the muscular fibre, contracting so powerfully under any known force. The analogy of the galvano-magnetic machine is only a very remote one, for the soft iron is from its nature already capable of becoming a magnet in other ways. In short, we are wholly ignorant of the intimate mechanism of the act of contraction, and cannot tell from that whether it be a vital or a physical act; for the

---

* As before said, the question of the dependence of muscular contraction on vital protoplasmic movements of the fibre itself, must be looked on rather as one not yet raised than as settled, and most physiologists seem to take it for granted. For instance, Dr. B. Sanderson at once concludes that the contraction of the leaf of the Dionæa muscipula is of the same nature as that of the muscles of vertebrates. He says also, that in muscular contraction an exactly similar change of shape in every particle of the muscle takes place; for "a muscle is not an apparatus made up of parts differing from each other in structure, but a mass of substance *equally instinct with life in every part.*" Also, that in the plant the agent of contraction is the protoplasm of the cells of the contractile organs, which, under stimuli, undergoes a most peculiar change of form and arrangement. With all respect due to so high an authority, I cannot accept the inference that because the contraction of the plant leaf depends on changes, apparently in the contents of the cells, the muscular contraction of the higher animals is of the same nature. By a parity of reasoning we should be compelled to conclude that the functions of respiration and digestion are not performed by the special apparatus provided for them in the higher animals, because these functions can be performed, in a kind of a way, by formless masses of sarcode. The causes of the movements and the locomotion of the Diatoms, Desmids, Bacteria, and flagelate Infusoria, and also the ciliary motions, are as yet unknown, though it may be allowed they are in some way dependent on the existence of protoplasm. And although a mechanism through which it can work in these organisms is not yet distinctly known, that forms no reason why a physical mechanism should not exist in the muscular fibre, nor that the latter should be "instinct with life in every part."

determination of that we are thrown back upon other reasons such as have been here detailed, and which show a preponderance in favour of the physical view, although we must be prepared for a possible reversal of this judgment in the progress of knowledge.

CHAPTER IX.

ON THE NATURE OF LIFE.

WE must now turn again to Fletcher, to see the protoplasmic theory of life in its purest form, for Beale, unfortunately, has obscured the question by the revival
of the vital principle as the efficient cause of the peculiar attributes of the protoplasm. I think, however,
it can be shown that only a very narrow line divides
the opinion of the two authors, and that in fact, practically, the theories of both are identical, as the supposed vital principle explains nothing, and is therefore
superfluous. At the same time, I think it is hardly
worth while taking into account the views of other
persons than these two authors on this question, because unless the generic unity of the protoplasm be
admitted the whole theory is untenable, and as yet
none but those two have adopted it in its full extent.
Fletcher, we remember, holds vitality to be synonymous
with irritability, or the faculty of undergoing upon
the application of a stimulus, any change not strictly
chemical or mechanical. This property resides in a
substance called by him "irritable," or "living"
matter, which undergoes corresponding changes with
every manifestation of "life" or "irritation," for "the

degree of irritability in any part is necessarily in the direct ratio of the quantity and quality of the irritable matter which it contains." Further, that this matter is "continually renewed and consumed by molecular processes" in all vital acts. And again, that this matter is in a state of combination so peculiar as to be absolutely sui generis, and on which none of the ordinary chemical agents and forces can act at all in the way they do on non-living matter; nor can they cause ordinary matter to pass into this peculiar state. That can only be done by pre-existing living matter, and moreover, this matter cannot pass back to the state of ordinary chemical combination, except by a process of its own peculiar nature, *i.e.,* its death as well as its birth is a specially vital process (iii. 144). *Life is,* therefore, not an entity, nor a force, but an action—and moreover, *that action* alone *which is involved in the consumption and regeneration,* from pabulum, *of* a material compound entirely sui generis called irritable matter or *protoplasm,* under certain conditions and stimuli. Just as combustion is not the materials for it, nor the product, nor the active force evolved, but simply the act of chemical union; nor does any supposed fire substance or element take part in the process. Just as little does any life substance or principle, material or immaterial, take part in vital processes. Our object is to reduce the idea of life to its simplest form in order to distinguish between the processes which are vital proper and those which are chemical and physical, all of which are engaged, more or less, in making up the functions of even the simplest order of individuality of living beings. The above definition approaches most nearly that of De Blain-ville, viz.: "Life is the twofold internal movement of composition and recomposition, at once general and continuous." While it is exempt from Mr. Herbert Spencer's objection that "it equally well describes the actions going on in a galvanic battery." From want of limitation to changes in one specially constituted substance, such as the protoplasm, I must object also to the definition of Mr. Spencer himself, viz.: "The definite combination of heterogeneous changes, both simultaneous and successive, in correspondence with external co-existences and sequences." Here is recognized, after John Brown and Fletcher, the fact that certain actions in correspondence with *external* agencies constitute life; but in the illustrations which follow, Mr. Spencer does not really distinguish between the chemical and the vital parts of the functions which are performed by living beings, although, as we shall see, he follows Fletcher in perceiving that they must contain a substance totally different from the chemical compounds known in the laboratory. With Fletcher, as with Beale, *all structure is dead*—the true living matter being always structureless—and inasmuch as structure is necessary to function, he distinguishes between that and action, defining a function as the action "of an apparatus destined to some specific purpose in the general economy of an organized being." Hence as all living beings perform certain functions by means of some non-living structure, however rudi-mentary, all definitions of life by the non-vital prin-ciplists which describe it in varied language as the sum of the functions, define really the life of individual

living beings, and not life in the abstract. This
applies, for example, to the otherwise excellent definition of G. H. Lewes. "Life is a series of definite
and successive changes, both of structure and composition, which take place within an individual without
destroying its identity."* Even Fletcher's definition
applies to individuals when he uses the term "organism" for the organized being as a whole; but not
when he defines life as "the sum of the actions of
organized beings resulting directly from their vitality
so acted on" [by pabulum, conditions, and stimuli].
Observe here the word *directly,* for it is only thus that
we can define life in the abstract, and separate what
is vital from non-vital in functions to which both kinds
of action contribute.† Fletcher uses the term organ-

---

* In his last work, "Problems of Life and Mind," published in 1874,
since he has apparently become acquainted with Fletcher and Beale,
although he does not mention them, he says: "The movements of the
bioplasm constitute vitality" (p. 118). In spite of the remarkable
clearness of thought and accuracy of expression in general of this
author, I must point out that this mode of speaking is not in agreement
with the definition of terms given by Fletcher at part ii. p. 5. The
term *vitality* can only be applied to the property or capacity of undergoing certain changes. What is meant is *life,* but it would not be
correct to speak of that as movement. The vitality of the bioplasm is
a certain state of composition which renders it capable of undergoing
certain *changes of composition* which constitute life. No doubt movements are essential to those changes, but they do not constitute life.

† For example, in a vertebrate animal, with circulating and digestive
organs, the resistance and support of the skeleton, and the tenacity
and elasticity of the blood-vessels, are physical attributes, but the formation and maintenance of these and all other tissues are vital. The
first stages of the digestion of food are mere chemical action, but the
absorption by the villi and conversion of it into blood, are vital. The
perception of stimuli is vital, but their transmission through the nerves
is physical. And so on through all functions. There is probably, or
rather certainly, no individual, however low in the scale, which has not
something physical and chemical in the "sum of its functions," besides
the distinctively vital.

ism also for that arrangement of matter into the
peculiar state explained in protoplasm, and with him
the word "organization" means the *process* by which
a being possessed of organs is formed—not the result of
the process.   With this distinction between action and
function, and organism and organization, we escape the
confusion into which Mr. Spencer frequently falls, as
well as Professor Huxley when he speaks of the rhizopoda exhibiting "life without organization"—this
last word meaning no doubt structure palpable to our
senses.   The truth is the so-called evolutionists like H.
Spencer and Häckel are perpetually hampered in their
definition of life by their desire to explain away the
unfathomable gulf fixed between vital and all chemical
actions, and thus to leave open a way for the origin of
life from natural chemical processes, while Fletcher—
a vitalist although not a vital principlist—as well as
Beale and all the vital principlists are content to leave
it unexplained.   Beale's objections * to the definition
of life as the sum of the action of living beings do not
apply to Fletcher, who had already limited life to the
action of the protoplasm alone.

The question now presses whether our previous
knowledge of the chemical properties of matter gives
us reason to suppose it possible that such remarkable
powers as those of life can be the attribute of any mere
combination of matter.   In addition to what has been
said on this point by Mr. H. Spencer and Dr. Bastian,
I have brought together † the chief facts in evidence
of the influence of not only chemical combination, but

* "Protoplasm," 3rd edit. p. 74.
† For part iii. of " Life and the Equivalence of Force."

mere complexity of atomic grouping in developing new and unexpected properties in compounds. This evidence is too voluminous to be given here, I will therefore only give a summary of the conclusions. The elements themselves are in all probability not simple bodies, their specific properties being given by the combination of atoms of still simpler bodies; these elements may exist in several different or allotropic states, displaying different properties owing to the mere grouping of their particles. When we come to bodies composed of several elements we have an almost infinite variety, increasing as might be expected with the number of separate constituents; but not only that, we have an additional element of variety in the mere complexity of grouping of the atoms of the same elements going to make up the molecule of the compound—as evidenced by the series of isomeric, polymeric, metameric, and homologous bodies. As an example, it is noticed that in twenty-seven volatile oils, including those of chamomile, hops, turpentine, clove, lemon, and valerian, the carbon and hydrogen are united in the same proportion, viz., ten to sixteen atoms. Another evidence is the remarkable physical properties of matter in the colloid state, which is attributed to the number and complexity of grouping of the atoms in the molecules. From these and similar considerations we are entitled to conclude that mere chemical combinations and reactions are sufficient to explain (when once formed) the properties of the infinite variety of tissues and fluids which make up the dead framework and secretions of organized bodies and even to account for some of the processes which go to make up their functions during life; and moreover that there is a *strong probability* that by still further complexity of composition and grouping matter may be capable of existing in the form of a substance possessing the properties of truly living matter. But that is the utmost we can say. I have reviewed the arguments of H. Spencer, Häckel, Bastian, and others, who bring forward as efficient causes of vital action, the instability and proneness to decomposition of the large moleculed organic colloids (Spencer); the faculty of imbibition (Häckel); the processes of catalysis and fermentation, and the fact that a number of chemical compounds usually formed by living bodies can also be made from organic elements in the laboratory. From these facts the conclusion is drawn by the school to which the above writers belong, that there is an unbroken chain of more and more complicated, merely chemical actions from the simplest reaction up to the complicated process of germ development. In opposition to this, reasons are given showing that the instability of the colloids has no counterpart in the living matter which has the power of self-renewal and maintaining its existence for years, and that, in fact, in this respect the living matter displays the strongest contrast rather than resemblance to the chemical colloids, for, as held by Fletcher, followed by Beale, the passage back to the chemical state or death of the living matter is a positive act belonging still to the vital series instead of mere negative chemical instability. That the faculty of imbibition no doubt is possessed as a physical attribute by the protoplasm as well as dead colloids, but that explains nothing of its peculiar

power of growth and self-renewal from heterogeneous
matter; that the process of catalysis does not really
resemble vital action at all, inasmuch as, in it, definite
forces tending to double decomposition are always present, and besides the catalytic agent never renews
itself; that although chemical fermentations may make
two and even three different products [*e.g.*, amygdalin
with emulsin and water forms bitter almond oil,
prussic acid and glucose] without consumption of
the ferment, and some of these of higher molecular
complexity; yet these ferments do not grow like the
organic ferments such as yeast and the putrefactive infusoria, and to argue from one to the other is simply
to beg the question by the poor verbal fallacy of calling two different things by the same name.

Finally, the extraordinary difference in the temperature and whole process of formation of organic compounds in the living organism, and in the laboratory,
give the strongest proof that some totally different
agency is at work instead of showing any analogy.
In addition to these the individual proximate principles of organic bodies * are shown to be destitute of
the properties of living matter. The conclusion is that
it is impossible to uphold the material theory of life
unless we admit a profound and ineradicable distinction
between the merely chemical and physical functions
performed by living beings and those which are vital,
and that the latter must reside in a substance, although
chemical, no doubt, in a wide sense—still so absolutely
different from matter in its ordinary state that it should
be put into a category by itself. Not to multiply

* Op. cit., §§ 68 to 75.

words I have suggested that this peculiar state of combination should be designated by Schwann's word, *metabolic,* which should be restricted to what pertains to it, and we may then speak of the metabolic state of matter, metabolic action, metabolic affinity, &c., instead of the ambiguous corresponding term vito-chemical.

Fletcher's hypothesis of the peculiar, vital, or metabolic state of matter for long met with little attention, being probably confounded with the speculations of Oken respecting the primordial slime (Urschleim), from which all living things were supposed to have arisen spontaneously, and gradually developed into the species—a speculation now brought into vogue again by H. Spencer and Häckel. Nevertheless, when other minds began to think in the same track, they could not fail to hit upon similar ideas, and accordingly we find Mr. Herbert Spencer, in his " Biology," published in 1864, puts forward a very similar idea in what he calls " Physiological Units." In the elaboration of this theory he follows, step by step, the same train of thought as Fletcher, arguing that the proximate chemical principles, or chemical units, " albumen, fibrin, gelatine, or the hypothetical protein substance," cannot possess the property of forming the endlessly varied structures of animal forms. Nor can any such power be given to the cell as a *morphological unit,* even if it had a right to that title. Therefore, he concludes, " There seems no alternative but to suppose that the chemical units combine into units immensely more complex than themselves—complex as they are; and that in each organism the physiological units pro-

duced by this further compounding of highly compound atoms have a more or less distinctive character.
We must conclude that in each case some slight difference of composition in these units, leading to some
slight difference in their natural play of forces, produces a difference in the form which the aggregate of
them assumes" (i. p. 182). We have here an indica-
tion of the great influence that may be exerted on the
ultimate varieties of power of the different kinds of
protoplasm by slight changes of composition of the
molecules. In the further development of the subject
he uses these physiological units chiefly to explain the
plastic faculty of living beings, viz., the innate tendency of the living particles to arrange themselves
into the shape of the organism to which they belong,
"just as in the atoms of a salt there dwells the intrinsic aptitude to crystallize in a particular way."
This, for want of a better term, he proposes should be
called organic polarity. By these same units he explains also heredity and variation. "Sperm cells and
germ cells are essentially nothing more than the
vehicles in which are contained small groups of the
physiological units in a fit state for obeying their pro-
clivity towards the structural arrangement of the
species they belong to" (254). "The likeness to either
parent is conveyed by the special tendencies of the
physiological units derived from that parent." I do
not suppose these statements are put forth as explanations, but their value lies in the admission that these
truly vital actions must belong to a substance possessed by living beings and vastly different from the
chemical proximate principles. Mr. Spencer, I per-

ceive with regret, does not mention Fletcher. Perhaps, as he is not a professed physiologist, he may not have heard of him, but what are we to say to the fact that he does not mention Beale, whose lectures had appeared four years before? It would have been better if he had studied both these authors, for his theory is at present most imperfect, and will bear no comparison with the complete protoplasmic theory of Fletcher and Beale. Indeed, it has the appearance of an afterthought, when, on reflection, he perceives that colloidality, instability, osmosis, and other chemical and physical properties to which he attributes a large portion of the functions of living beings, hastily named by him vital, are wholly unable to throw any light on the really and distinctively vital actions. The imperfection of the theory is shown on attempting to adapt it to the phenomena of life on frequent occasions. For instance, in attempting to explain evolution, he says (vol. ii. p. 11): "Such being the primitive physiological units, organic evolution must begin with the formation of a minute aggregate of them—an aggregate showing vitality only by a higher degree of that readiness to change its form of aggregation which colloidal matter in general displays, and by its ability to unite the nitrogenous molecules it meets with into complex molecules like those of which it is composed." Now, granting that we have got our complex units, see here how the grand characteristic of living matter is slipped in as a thing of little consequence, and quite on a par with the colloidal instability supposed to belong to living matter, but does not. This faculty of combining heterogeneous compounds into matter like itself—growth, in fact—is the very thing possessed by no other substance in the world, and which is possessed by all living matter. He has thus here in words leaped across the unfathomable gulf that separates living from dead matter without recognizing it; and until the difficulty is fairly met, and all vital action referred to some peculiar complex substance in a state of generic unity though capable of infinite specific variety, while all structure and physical and chemical action belong to matter in its ordinary state, the materialist theory of life will be untenable. In short, the physiological units must be all in all for life, or the supposition of them is of no use at all. By other physiologists the existence of different and more complex compounds than the proximate principles are spoken of, but in a vague manner; by Häckel and Huxley, as we shall see in a subsequent chapter, but not with the distinct line of demarcation above given. Among others, Ranke says, " The chief constituents of the protoplasm seem to be albuminates in a state of imbibition with water, or *still more highly composite stuffs*, which, like hæmoglobin and vitellin, give rise to albuminates on their decomposition " (p. 80).

It may be interesting to say a few words on the possible physical state of the living matter. The foregoing considerations of the development of properties by complexity of constitution have been summed up by Dr. H. Madden, in his brilliant essay, " On the Relation of Therapeutics to Modern Medicine," 1871. Dr. Madden adopts my suggestion of restricting Schwann's word metabolic to designating the atomic constitution of the living matter, and my conclusions

respecting the nature of force and the distinction between force and property, which reconcile Fletcher's
hypothesis with the modern doctrine of force and
bring it into harmony with Beale's discovery of the
unity of the protoplasm.   He then goes on to say,
" We have now, therefore, reached the confines of
physiology, and we have seen four classes of complexity, each of which contains many degrees, viz.,
1st. Atomic complexity, varying in accordance with
the atomic weight of the element; 2nd. Chemical complexity, varying according to the number of elements
and the number of atoms of each which go to form
each chemical combination; 3rd. Colloid complexity;
4th. Metabolic complexity.   We have likewise seen
that this last and most complex molecular arrangement of all must be reached ere the phenomena of life
can be manifested."   While agreeing with this scale
of complexity as a matter of fact, I would state it differently, in order to retain the simplicity of the sharp
antithesis between chemical on the one hand and
metabolic on the other.   So, putting aside for the
present the compound nature of the elements, and as
there is no question but that the colloids are simply
chemical compounds, we have the reactions of the elements up to the degree of colloids and all organic
proximate principles, called *chemical*, while those
higher reactions characterized by life are called *metabolic*.   When we call to mind the hypothesis of
Samuel Brown, that the elements themselves may
owe their distinctive properties to the mere number
and grouping of certain ultimate atoms of one kind—
a hypothesis now almost proved true by the more

recent discovery of spectrum analysis*—this lends
additional probability to the development of a state of

* The bearing of the spectrum analysis on this question is the subject of several admirable mathematical papers by Mr. Ponton, in the
"Quarterly Journal of Science," 1871, from which the following
points are extracted.   If the elements had been simple, homogeneous
masses, of definite size and weight, each element in the state of incan-
descent vapour would have exhibited in the spectrum one single bright
line; but this is, we know, not the case, and even hydrogen, the lowest
in the scale of chemical equivalents, shows four bright spectral lines,
while iron presents a large number.   Hence the great probability is,
that "each ultimate particle of the element consists of numerous more
minute atoms, differing in their inertia, and held together by a force
too great to be overcome by any chemical means which can be brought
to bear upon them."   The smallest particles of the chemical so-called
elements he proposes to call "ultimates," and the word atom is reserved for the still smaller constituents of these.   Thus "molecules"
will denote the particles of chemical compounds, and ultimates and
atoms as above.   The atoms constituting the chemical ultimates, while
very close to each other within the limits of the ultimate, cannot be in
absolute contact, otherwise they would be incapable of separate individual vibration.   And the circumstance that the ultimates absorb
definite lines in diverse and distinct parts of the spectrum shows that
the motion must be taken up by different atoms, having, in virtue of
their intrinsic inertia, a tendency to vibrate at those different definite
rates.   From the calculations given, he concludes that the ultimate of
hydrogen is made up of a great number of atoms of four diverse sorts,
the number of atoms of those sorts being different.   A similar complex constitution obtains with all the other elements to a greater extent than with hydrogen, and it is concluded that several of the atoms
are common to the ultimates of several different elements.   Thus the
species of atom which in the ultimate of hydrogen produces the line
C, occurs also in the ultimates of six metals: several atoms are common to iron and titanium, to iron and calcium, to iron and nickel, &c.
No force is known on our planet, or even in our sun, which can tear
asunder the constituent atoms of the ultimates, but it is thought that
in the more intense heat of Sirius, which is said to be a younger star
than our sun, the atoms are dissociated, and the elements are thus not
yet formed.
    All this is substantially confirmatory of the views of Samuel Brown
on the nature of the so-called elements, as laid down in his lectures on
the Atomic Theory in 1843.   His position was that the so-called atoms
of the elements [ultimates of Ponton] are "made up of homœomeric
parts, not essentially indivisible, but indivisible by such forces as are
competent to the division of their aggregates."   And again, "Atoms
are defined as indivisible by such forces as divide their aggregates or
the forms produced by their concourse."   Almost the same statement
on the divisibility of matter and definition of a molecule are given by

matter such as the metabolic by complexity of combination and grouping.

The supposition of three distinct states of combination of matter, all comprised in the term chemical in its wide sense, but which we have no power of interchanging at will in the laboratory, is not unphilosophical. Almost all systematic treatises on chemistry describe the elements as bodies which have not yet been decompounded, though not necessarily undecompoundable; while no one asserts the practicability of their transmutation. On the contrary, the whole fabric of modern chemistry rests on the certainty that you will get back, weight for weight, every particle of those elements on analysis: and never that you may get weight for weight of the atoms of one element combined into another or split into several other ele-

---

Prof. Clerk-Maxwell, in his text-book on "Heat," in 1872. Means being thus wanting to decompose the elements, S. Brown believed, however, that he had demonstrated their compound nature by synthesis, and that he had formed silicon by uniting four atoms of carbon. These experiments have not succeeded in the hands of any other chemist as yet, and therefore we must suppose there must have been some undetected source of fallacy, either in his experiments or the repetition of them.

We have seen that Fletcher held the living matter to be in a quite different state from ordinary chemical combination, and possessing powers of analysis and synthesis of which the experience of the laboratory can give us no idea. And from the difficulty still existing of accounting for the origin of some elements in certain plants, he was inclined to the belief that the elements themselves were decomposed, and re-combined in the living matter. Samuel Brown, in accordance with his theory of the compound nature of the elements, agrees to the probability of the supposition, and says, "Indeed, the carbon, oxygen, hydrogen, and nitrogen which have been extracted from organic bodies may often be mere products of transformation" (76). And he suggests that the silica found in the larger reeds, as the bamboo, grown in soil without an apparently adequate supply of that earth, to the puzzlement of botanists, may be formed from carbon by the living plant. It is proper, however, to say that an absolute deficiency of sufficient silica in the soil has not been demonstrated.

ments. And yet strong reasons are not wanting for the belief that these elements are made up of certain ultimate component atoms, held together by a strength we cannot yet overcome. In contradistinction to those simplest compounds, over which chemistry has no power of analysis, it is not unreasonable we should suppose compounds of the opposite extreme of complexity, over which we have no power of synthesis and no power at all chemically except for destruction. The range of what may be properly called chemical action must be held to extend from the elements up to the protoplasm (not inclusive), thus comprising what is now called organic as well as inorganic chemistry, while it has no place in the vital, or metabolic, or protoplasmic state. The molecular actions of the protoplasm constitute, in fact, physiology.

To all arguments against the vital or metabolic state being a mere material combination, without any new vital power or principle added to it, on the ground that we cannot set it together by any chemical process, we point to another truly material combination, viz., the elements, which we have just as little power to unmake. Besides there are many chemical compounds, such as the diamond, and other precious stones, and, as yet the vast majority of organic chemical compounds, which we can destroy but cannot yet construct, although nobody thinks of attributing their properties to any spiritual principle. But between the chemical and the metabolic states we are justified in believing there is a greater gulf fixed than between transmutation of the elements on the one hand, and ordinary chemical reactions on the other.

With respect to the nature of the metabolic state little can be said. Fletcher speaks of it as "an ever-varying form of existence,"* and Dr. Madden illustrates this by analogy with well-known facts in chemistry. The ordinary chemical affinities are suspended, he says, because of the perpetual change going on.

"Most chemical combinations require time, and the rate at which the constituents are brought together, and the duration of their contact, will materially influence the result; so much so, indeed, that advantage is taken of this fact in the arts to accomplish purposes which would otherwise be unattainable. For example, before calico can be printed, every loose particle of cotton must be removed from the surface, in order that the coloured inks may not run, and damage the clear outline of the pattern. This removal is effected by passing the calico over, and in contact with, a red-hot iron cylinder, and by regulating the rapidity with which the cylinder revolves and the calico passes over it, the intense heat burns off the loose fibres and yet does no injury to the woven cloth. In other words, the changes in the relation of the high temperature and the cotton are too rapid to admit of the fibre combining with the oxygen. Let the rate of revolution be reduced but a very little and the calico would burst into flames. Again, it has been found that certain fulminates can be detonated in contact with gun-cotton without causing the latter to explode; and experts account for this on the ground of the extreme rapidity with which these fulminates expand, too rapid, indeed, to enable the pyroxyline to initiate its new mode of motion, and hence it remains unchanged. Precisely the same kind of thing occurs in the metabolic state of matter. It can only last so long as rapid and incessant changes are going on, for which purpose it must

* And again, "The really organic elements or molecules are probably, under ordinary circumstances—that is to say, while the organized being not only possesses the aptitude for life, but manifests life itself—never for one instant the same, and are certainly such as to have entirely eluded hitherto all our attempts to overtake them."—Fletcher's "Physiology," p. 133.

always be in contact with pabulum; and if the rate of these changes is reduced beyond a certain point, the chemical affinities of the materials will at once assert themselves, and the whole will break down into more or less stable chemical combinations" (p. 5).

This may be: but we must remember that the living matter has also long intervals of a resting stage, or even a state of suspended animation. In seeds, also, and germs, life seems dormant for long periods, and all change is reduced to a minimum. What that minimum may be we know not, for the rest is never absolute, and Dr. Beale remarks that some change, however slow, must be going on as long as life lasts. I do not attach much weight to any speculative opinions on these points.

With respect to the physical condition of the living matter, the molecules themselves must be constituted by atomic combination into the vital or metabolic state, but no part or organ, or probably even single plastid, is composed of molecules of only one kind, as is the case with crystals. Hence the different molecules which make up every mass having its specific vital attributes, must be connected by cohesive force, although slight; and it is a fundamental fact that life does not exist in matter in the gaseous or liquid, but exclusively in the solid state. The apparent exception of the blood, lymph, and other fluids is easily explained by the presence of floating masses of living solids in them. But this solid state is very peculiar and one that admits of the interpenetration of a large quantity of water, without dissociation of the molecules into that free and mobile state which constitutes liquidity.

At the same time, sufficient mobility is given to facili-
tate those rapid metabolic changes, and the access of
pabulum and the removal of products which are essential to vital action. For this reason, also, the living
matter can never exist as a rigid structure of any
kind. This peculiar semi-solid state is called the state
of imbibition, and is so remarkable that Schwann
gives it a prominent place in the process of building
up the organic structures. In fact he is inclined to
the opinion that "organisms are nothing but the form
under which substances capable of imbibition crystallize" (215). And Häckel, following Schwann, suggests
that the imbibition state should be considered a fourth
physical state of matter,* as it is to be distinguished
from the mere humidation of aggregations like sand-
stone ; and the fact that crystalline bodies are incapable
of imbibition is one reason why they are incapable of
living action. He exalts this faculty to the skies as a
main factor in the production of vital phenomena, by
bringing the pabulum into the interior of the mass,
contrary to what happens in crystals. But in reality
it explains nothing at all, for when the nutriment has
got into the interstices of the protoplasm the difficulty of explaining the peculiar changes it undergoes
there, only begins. In fact this is not a distinctively
vital state at all, but belongs to ordinary colloids.
But there is a difference between the physical state of
the colloids and that of the living matter, for the
former are, in many instances, capable of true solution,
e.g., gelatin, which no living matter is. The tearing
asunder of the molecular groups by the adhesive force

* "Generelle Morphologie," i. p. 124.

of the particles of the solvent, produces decomposition
of the complex and delicate living molecules and their
death, just as mechanical force of any other kind, such
as pressure, blows, tearing, &c. Although a large proportion of water must be contained by imbibition in
the living matter, yet a certain proportion cannot be
exceeded without injury. Accordingly, the living
matter is always protected from the solvent action of
pure water either by cell walls or by being kept
bathed by fluids not so strongly diffusible, as is always
the case in the higher animals. Pure distilled water,
in fact, is rapidly fatal to protoplasm when placed in
contact with it,* inducing death and consequent coagulation. In the lowest tribes, which are chiefly aquatic,
such as the infusoria, when they consist of apparently
nothing but a naked lump of protoplasm, no doubt,
this substance is so constituted as to repel excess of
water and prevent diffusion and solution. No comparison with the physical state of dead matter can give
a correct idea of that of living matter as we see it in
the amœbæ and other rhizopoda. The slow crawling
movement, the rapid pouring out of the substance in one
direction, and, again, the gathering up of the whole
mass into a ball, in short, the numberless changes of
shape without change of volume, show a mobility
almost equal to liquids, with a cohesive power like a
solid, and in fine physical properties, possessed by no
other substance. This very peculiarity of the living
matter viz., that it is always semifluid, transparent

* "The plasma usually appears as a semi-fluid, albuminous body,
of the consistence of a tough, sticky, thread-drawing mucilage, which
is insoluble in water, and even, in many cases, coagulates by the access
of water" (Häckel, "Gen. Morphol.," i. 278).

13

colourless, and perfectly structureless, and therefore
one mass is quite indistinguishable in aspect from
another of a different kind, is made the groundwork of
Dr. Beale's objection to the materialist theory. He
says over and over again, how can such a variety of character and action, the self-moving, the dividing, and the
forming structures unlike itself and of infinite variety,
depend upon the mere arrangement of the atoms in
this apparently simple, uniform substance ? He objects
particularly to the term "molecular machinery," which
he seems never tired of ridiculing, and seems to think
that our mere inability to understand or to express,
except by figurative language, the intimate nature of
the action in the protoplasm gives, somehow, support
to the hypothesis of an immaterial principle added to
the protoplasm.   I fail to see the cogency of the argument, and rather think he has added to the difficulty
by a suggestion which stands in more need of explanation than the thing it purports to explain.   The truth
is, the apparent uniformity of the protoplasm in outward appearance, while capable of infinite variety of
inward composition, is nothing more than is found in
all large classes of chemical compounds.   Who expects
all transparent and colourless liquids to have the same
composition ? and so on, with fifty illustrations.   The
infinite variety of faculties of the different kinds of
living matter was, of course, at once attributed by
Fletcher to a corresponding variety of molecular composition, and no one since, except Dr. Beale, has found
any difficulty in doing the same.   This variety of
composition is included in Fletcher's term " organized,"
and has been more happily termed " molecular organiza-

tion" by Dr. Bastian. The possibility of the sufficient variety of the said molecular organization is well expressed by Gegenbauer, who says that the complicated formal life-phenomena of the protoplasm—even granting that it cannot be analyzed farther anatomically—are nevertheless of such a nature that they not only pre-suppose a more complicated molecular constitution than we can as yet understand or imagine, but that the protoplasm may in this respect be placed on an equality with complicated organized beings; and Professor Rutherford thus expresses himself, "There appears to be no reason for supposing that two particles of protoplasm, which possess a similar microscopic structure, must act in the same way; for the physicist knows that molecular structure and action are beyond the ken of the microscopist, and that within apparently homogeneous jelly-like particles of protoplasm there may be differences of molecular constitution and arrangement, which determine widely different properties" ("Brit. Assoc.," 1873).

The theory of the constitution of chemical compounds of S. Brown, taken in connection with the recent revival of dynamic theory of molecular action of Lucretius by the mathematical physicists, is most interesting to us, especially as bearing on the supposed spontaneity of vital action. It is becoming rapidly accepted among physicists that, instead of being at rest, the particles of matter are all in a state of vivid molecular movement, and by this are to be explained nearly all the secondary properties of matter. The elasticity of the gases and the diffusion of liquids are already proved to depend on mere molecular movements; the theory agrees also with the doctrine of interior work, and explains the apparently spontaneous action of chemical changes in solutions, and although the exact

molecular state which produces the rigidity of solids, is not
yet known, yet the molecules are not there at rest, being agi-
tated by heat and, probably, intra-molecular movements. The
hypothesis of S. Brown here comes in. He holds that the
chemical state forms always a counterpart of the astronomical
state of the distribution of matter. That in each molecule
formed by the combination of separate atoms we have, as it
were, a solar system. The atoms are supposed not to be indefinitely near each other, as was uniformly assumed, but their
"distances may be as great in proportion to the diameters of
those particles as, say, the distances of the planets from the
sun, in proportion to their diameter and his."* In the molecule
the "atoms revolve around each other in the line of the revolution of the centripetal and centrifugal forces, while of two
unequal and dissimilar particles (as an oxygen and a hydrogen)
the less particle shall be planetary and the other solar." "Each
water-particle is a true unit, with its centre neither in the oxygen particle nor in the hydrogen, but in the shifting focal point
of the forces of both." In commenting on Dalton's reason why
carbon cannot combine with more than two atoms of oxygen,
viz., that "in the state of CO_2 there are two atoms of oxygen
combined with one of carbon, and a third or fourth atom of
oxygen, however it may be attracted by the carbon, cannot join
it without repelling one or more of the atoms already combined.
The attraction of the carbon is able to restrain the mutual repulsion of two atoms of oxygen, but not of three or more." To
this S. Brown says, "This lucid conception may be made more
lucid still, perhaps, by the counter statement in astronomy
that a sun cannot be overloaded with planets." Then, after
illustrating this by the constitution of ammonia, he concludes,
"It remains to be seen whether there are any data of a purely
dynamical sort to determine *how many* planetary hydrogens
there must be in a compound atomic system of ammonia. If
there be, or if any can be found, then the atomic theory shall

* Professor Challis also concludes that the proportion of space occupied by the atoms is exceedingly small compared to the ethereal streams
and movements surrounding them ("On the Math. Principles of
Physics," 1873).

be perfect." Again, in illustrating chemical decomposition by the instance of potassa and carbonate of lime, whereby the carbonic acid cleaves to the potassa and quicklime is set free, he says, "It is precisely as if some stronger planet were brought near enough to draw the moon off from the earth ; in which case the compound unit called the terrestrial system, composed of the earth and the moon, would be decomposed." It is obvious how this harmonizes with the more recently found law of atomic values or atomicities, and I do not know that any other even plausible physical theory of it has been formed. The diagrammatic representation of the position of the atoms given in the form of tooth-combs, balls on rayed spikes, bricks, &c., seem to be contrived as if to show how they could not possibly exist in nature. It also gives us an idea how force may be accumulated within the molecules in the form of interior work, as it is known to be. To return to the state of living matter. "Let us," says Beale, "in imagination peer into the ultimate particles of the living, active, moving matter, and consider what we should probably discover. Were it possible to see things so very small, I think we should discover spherules of extreme minuteness, each being composed of still smaller spherules, and these spherules infinitely minute. Such spherules would have upon their surface a small quantity of matter differing in properties from that in the interior, but so soft and diffluent that the particles might come into very close proximity. In each little spherule the matter would be in active movement, and new minute spherules would be springing into being in its central part. Those spherules already formed would be making their way outwards, so as to give place to new ones, which continually arise in the centre of every one of those animated particles. . . . . The change which occurs in the living centre is probably sudden and abrupt. The life flashes, as it were, into the inanimate particles and they *live*" ("Prot.," 3rd ed., 277).

In this smallest conceivable mass there exists, as Beale frequently repeats, fluid pabulum passing into the centre, being there transformed, and working outwards till on the surface a portion dies into the formed material. Here are then conditions for action independent of spontaneity, and the mere circumstance of vivid molecular movement can be paralleled in all liquids, especially where chemical changes are going on. To peer then farther in imagination into the constitution of those spherules, we may picture to ourselves the very elements resolved into their component atoms, and these, rearranged into clusters and constellations of molecules of relatively enormous size and complexity. These latter, again, representing systems made up of atoms corresponding to suns, planets, and satellites, all revolving rapidly in their orbits, while possibly also, as supposed by Williamson in solutions of salts, some atoms may be passing from one molecule to another. In this state of matter, so different from the simpler inorganic state, we may suppose new laws of vito-chemical or metabolic combination. In all *chemical* combination double decomposition and recombination with formation of a third substance into which both factors enter in exact proportion, are the rule, while here, when a chemical substance—pabulum—is placed in contact with the living matter, the latter is not decomposed but the pabulum is instantly taken to pieces—possibly its ultimates even resolved —and incorporated with the living matter, from which it emerges in a totally different state of chemical combination partly as tissue or secretion, and partly as effete products, while any surplus force comes out as heat or mechanical movement.

Leaving what is merely speculative on the constitution of the living molecules, it is obvious that between ordinary matter and such a kind of material composed as protoplasm is, there can be no possible relation of a chemical kind at all ; and the only relations the living matter can have with matter in the ordinary state and with force are those of pabulum, conditions, and stimuli. To the living matter there are no acids nor alkalies, no solvents, no astringents, no fats or soaps, no ferments or catalytic agents, no sugars nor alcohols, no albumen, fibrin, gelatine or the like to act upon it in any way, resembling in the least their action on dead matter, but only as pabulum and stimuli. Probably moisture alone acts purely as a condition, but nearly all kinds of pabulum, including oxygen, act also as stimuli, while some agents generally recognized as stimuli, act in that capacity alone, while some are conditions as well as stimuli. Heat, for example, is an essential condition for all living action, and it is likewise a stimulus to all living matter, and if raised above certain limits, exalts living action, causing the organism to live faster—an exaltation soon succeeded by a corresponding depression. Electricity is also a stimulus to all living matter, but not, as far as we know, an essential condition, while light is a stimulus to only certain forms of living matter ; and magnetism hardly to any, as far as we know. All stimuli, comprehending all positive agents which act at all on the living matter, exalt the living action in degree, and this is followed by a corresponding depression if beyond the normal amount, and if the excess of action caused by normal stimuli is great or long continued, qualitative change as well as exhaustion of the living matter is induced. But with the preternatural stimuli, such as medicines and poisons, the depression after exaltation is more marked and also the qualitative change, so that Trousseau calls them modificators. The depression and modification is often so marked as to be the only thing to attract attention, and a direct paralyzing effect is assumed by many authors, but Claude Bernard, as the result of his long series of experiments, states that, " Every substance which in large doses abolishes the property of an organic element, stimulates it if given in small ones " (" Introduction à la Médicine Experimentale "). The preternatural stimuli no doubt become incorporated with the living matter in the same way as the pabulum and form part of it, till decomposed or expelled, and during the process produce the stimulation and modification of qualitative action which we recognize as the specific medicinal or poisonous effect.

Apparent Spontaneity of Vital Action.—If life is thus merely the action of a complex material compound, how are we to explain the spontaneous movements and changes it apparently undergoes ? No matter how complex the protoplasmic molecule may be, its atoms are still nothing but matter and must share its properties for good or evil, and among the rest, *inertia.* Hence it cannot change its state of motion nor rest without the influence of some force from without. True spontaneity of movement is, therefore, just as impossible to it as to what we call dead matter.

Nor if we look upon it as, what indeed it is, a compound of high potential energy capable of evolving force and thus moving itself by the rearrangement of its constituent atoms, does that help the question ; for the possibly very small initial force required for that change is just as impossible to conceive without adequate cause, as the whole force for moving it and keeping it moving for ever without equivalent communication of force. Some have suggested that the revival of the theory of Lucretius as to the dynamic state of all matter above spoken of, may help us here. But it helps nothing, for these move-ments are all in equilibrium, and to *change* the motion of a body already in motion requires an adequate cause just as much as to start it from rest. So we are compelled to admit the existence of an exciting cause in the form of some force from without, to give the initial impulse in all vital actions ; this is the stimulus. When this is once given the amount of action or work depends on the potential energy of the protoplasm and pabulum which are decomposed and may bear no calculable proportion to the initial impulse, just as a spark may explode any mass of gunpowder, or the flutter of a bird's wing start an avalanche. We thus see how the stimuli apparently produce results far beyond the intrinsic energy conveyed by them. It is on the whole question of stimuli that Dr. Beale differs so widely from other physiologists, and especially from Fletcher, who is, above all others, careful to keep the essentiality of them in due prominence. Dr. Beale does not even represent cor-rectly the position of the stimulus. He says* that if you call the development of the egg a consequence of the action of heat as a stimulus, you may as well say that a steam-engine is a con-sequence of the coal that takes part in generating the steam that turns the lathes that are used in its construction. This is not a correct statement of the position that the stimulus gives merely the initial impulse to a substance like protoplasm, which has not only a store of potential energy, but the *directing power* of using it by virtue of its marvellous molecular organization. The true analogy is with a steam-engine having its boiler charged, and the stimulus is, as it were, the power which turns

* " Protoplasm," 3rd edition, p. 270.

the cock and admits the steam into the cylinder. This necessity for an initial impulse, or the removal of a hindrance to
action is of immense significance in a teleological point of view,
for without it no living creature above the very lowest order of
individuality could exist ; no function requiring the harmonious
working of different parts of an apparatus could be performed,
and the whole organism would rush to its destruction by the
simultaneous over-activity of all its parts. Hence, in the building
up of the higher orders of animals there is a gradual differentiation of the living matter, so that each kind becomes susceptible
of full activity only under its proper stimulus. No doubt the
essentiality of the stimuli is often lost sight of because they
form part of other essential conditions of life, *e.g.*, pabulum,
oxygen, &c. ; but when more closely looked into, the action as
stimuli will also be found. Even when no visible apparatus is
present, the necessity for stimuli is recognized by most physiologists. Of the movements of the amœba and white blood
corpuscles Kühne says they take place from stimuli as yet
unknown. Dr. Beale, however, still clings to his theory of a
hyper-physical cause or vital principle which can initiate these
and other vital movements. He is, however, now greatly em-
barrassed. Formerly he did not hesitate to say they were
directly the offspring of a hyper-physical vital power, but
having stated that he did not assert the creation of force by
living matter, he is now obliged to deal with the subject in a
more guarded manner, and it is not easy to say what he does
mean in his third edition of " Protoplasm," where he says that
the antecedent change that occurred just before the vital movement cannot be proved to be *phenomenal*. While he says, " it
is scarcely incorrect to say that it moves of itself, because at
this time no one can adequately explain the cause of the movement " (269). He will not now say anything more distinct than
that he attributes these movements to " a *peculiar power* of
movement or to *vitality*," but the tendency of the remarks at
p. 273 is that they are spontaneous, and depend on no physical
cause. This is, of course, tantamount to a creation of force.
One must be, however, far more surprised that Professor Bain
upholds the doctrine of spontaneity of vital actions in his

classical work on the "Senses and Intellect." At p. 64, *et seq.*, he maintains, contrary to Fletcher, that a spontaneous Energy resides in the nerve centres which gives them the power of initiating molecular movements "without any antecedent sensation from without, or emotion from within, or any antecedent state of feeling whatsoever, or any stimulus extraneous to the moving apparatus itself." And this " fact of spontaneous activity I look upon as an essential prelude to voluntary power " (296). His chief proofs of this are the tonicity of the muscles ; and to the objection that cutting the sensory roots destroys the tonicity, he says, granting that it is really reflex and arising from perennial irritation of the incarrying fibres, that it is not what we usually understand by stimulation. This last he wishes to restrict to the effect of " visible and remitted applications to the parts. A constant stimulus is no stimulus at all " (64). This may, to a certain extent, apply to conscious sensation where relativity plays an important part, but it is quite inapplicable to mere irritation ; and the permanent action of stimuli, such as heat for example, is essential to continuance of vital activity. Nor would the teleological reason for the stimuli be satisfied by a mere initiatory impulse which will allow action to continue spontaneously till the whole is consumed, as in firing gunpowder for example. On the contrary, it must be, and is in most cases, a continued impulse under which the vital activity only takes place as long as it lasts. His other proofs are the contraction of the sphinctors and of the capillaries, the rotatory movements produced by uni-lateral section of the pons varolii, the activity of the involuntary movements, the initial beat of the heart, &c., all of which may be easily answered in the same way. But his final and, he thinks, most conclusive proof, is the priority of movement to sensation in wakening from sleep, a circumstance noticed by Aristotle, who referred these movements to an internal source. Bain holds that the refreshment of the nerve centres after repose causes a " burst of spontaneous energy" and " a rush of nervous power to the muscles, *followed* by the exposure of the senses to the influence of the outer world"(66). The facts can be, I think, interpreted differently and more consistently with general physiological laws. All life, with Fletcher, implies the consumption of protoplasm under the operation of stimuli, but this, especially in the activity of the nervous system, does not always go on *pari passu* with the regeneration of the life-substance ; on the contrary, it exceeds it at times ; hence, exhaustion and modification of living matter, so that the same stimulus loses its effect and can no longer rouse to activity as before. The living matter now passes into the resting stage and sleep comes on, during which the regeneration overbalances the consumption, and in due time the susceptibility is restored to its previous acuteness, so that the *same* stimuli, which are still present, are now sufficient to excite the nerve centres into the waking state of activity. Of course it is not pretended here to explain the nature of sleep, or of the resting stage of living matter as contrasted with its special activity. But even without the above connection of the phenomenon in question with consumption and regeneration of protoplasm, there is no doubt that the repetition of every stimulus produces enfeebled or exhausted susceptibility towards itself, which susceptibility is revived again after rest. In this sense it may be said, for a time, in respect to some stimuli, " a constant stimulus is no stimulus at all." In respect also to the storing up of the living matter, and even that chemical substance supposed to be consumed immediately in nerve and muscle action (see p. 159), the facts agree with Bain's view of the exuberant movements of the young and the vigour and freshness after rest, only the discharge of energy is not spontaneous. In fact it is as if the boiler of an engine were well stored with steam you would have to turn the cock all the same, although the machine would go faster and farther than when half cold and exhausted.

But the strongest proof of the non-existence of spontaneity is given by Beale's absolute denial of the existence of uni-polar nerve cells. There are always two poles at least, and often many more. If the nerve cell could start into activity spontaneously, and if any purpose were to be served thereby, one cord would be enough to convey the evolved energy as far as we know, for we do not know that the second is necessary for any electrical reason, *i.e.*, to complete the circuit. The second is no doubt an afferent or commissural fibre, whereby the stimulus is conveyed *to* the nerve cell for the purpose of exciting it to activity at the right time, and if so there can be, in health, no other mode of activity possible, for that would frustrate the very purpose of the organ. The intricacy of the nerve-cord connections of the bioplasts of the brain and their extraordinary number are dwelt on by Dr. Beale ("Protoplasm," 3rd edit.) : "A portion of the gray matter upon the surface of a convolution not larger than the head of a very small pin will contain portions of many thousands of nerve fibres, the distal ramifications of which may be in distant and different parts of the body." " The bioplasts referred to are directly concerned in mental action. Movement affecting the matter of many thousands of these minute bio-plasts, probably at the same moment, is required for the initiation of the simplest idea" (321). To produce, therefore, anything like harmony and coherence in ideas and movements, all these separate bioplasts or sources of energy must be rigidly tied down, never to begin to act except on their appropriate stimulus. Conceive the effect of any spontaneity ! It would be incoherence and delirium, speedily followed by destruction of the individual.

The essentiality of stimuli for all vital action, even in organic life, is not so obvious at first sight because the stimuli co-exist with pabulum and conditions in most instances, but it could be shown if it did not take us too far from our present subject.

For the same reasons as above, the spontaneity of the will must also be denied. To allow it any initiating power not derived from the transformation of pre-existing force, even to the small extent required by J. Herschel, would be to create force. I have no pretensions to attempt to unravel the intricacies of this most delicate and complicated subject, so I will only say that no doubt the perception of internal stimuli is here the initiating agency.*

* I mean, of course, the spontaneity of the will in its physiological sense. I have no opinions on the subject of the moral freedom of the will and moral responsibility different from those held by common consent in the age and state of society in which we are living, and which admit that qualified amount of freedom of the will which common sense shows we are possessed of.

CHAPTER X.

NOTHING has, I think, of late hindered us from forming a clear conception of the nature of life, so much as the way in which the relation of the physical forces to the action of living beings has been viewed. And this has mainly arisen from a want of due care in distinguishing force from property, and from an exaggerated conception of the office of the former which is virtually by some exalted to the rank of a semi-intelligent power, *vice* the vital principle deposed. The fact is, that many medical writers, not having gone to the root of the matter for themselves but only accepting the current definitions, have been unable to resist "the insuperable tendency of the human mind to personify its abstractions," and have taken up the notion that force, or energy, is a mysterious something capable of existence *per se*, and possessing various positive, and even intelligent (such as *formative*) powers. This is quite wrong. Force is not an object capable of existing *per se*, but is only the motion of matter and æther, or the pressures antecedent to that motion, whichever way you choose to apply the word, but, at all events, it is never anything but an affection of matter, including the æther. To speak of points or lines of force as the physical cause of cosmical phenomena without any substance, which is not force, underlying them, is inadmissible, and seems as repugnant to common sense as the idea of motion without something to be moved.* There is no *mystery* about force in the sense of its being above and beyond our intelligence, like the things pertaining to the spiritual world, but there is much that is yet hidden till we get a clearer knowledge of the physical natures of the æther and of the atoms of matter.

* " All the least parts of all bodies are extended, hard, impenetrable, and indued with *vis inertiæ.*" Challis quotes this passage from Newton and adopts it, as likewise Newton's theory of an æther which, however subtle, is composed, like air, of parts which mutually press. The idea of an atom as an extended body, with infinite hardness and strength, is not now entertained by physicists in general, but still, as stated by W. Thomson, " we must realize it as a piece of matter of measurable dimensions, with shape, motion, and laws of action, intelligible subjects of scientific investigation " (" Brit. Assoc.," 1871). Helmholtz's theory of vortex-motions in an incompressible frictionless fluid is now looked to most hopefully, as likely to give the true idea of the nature of the atoms : but this does not countenance the notion of self-existent force, for the æther itself is material. Mr. Macvicar, in his original and profound "Sketch of Philosophy," No. II., postulates the æther as consisting of discrete spherical particles, all equal; and that the material elements are built up of clusters of these, forming units, and again combinations of these units into certain stable shapes, forming the elemental atoms.   Dr. Samuel Brown favours Boscovich's well-known hypothesis of points of repulsive and attractive forces, and of these he (S. B.) supposes five spheres of alternate attractions and repulsions, which are since nearly paralleled by Challis's series of æthereal waves of attraction and repulsion.   But when pressed to explain the nature of this ultimate repulsion without substance, S. Brown used to give utterance to the following sublime conception, which I cannot refrain from repeating, although I do not find it in his published works. The ultimate repulsion constituting the extension and impenetrability of the atoms of matter, could be conceived of in no other way than as the persistent exercise of the Will of God Himself, " in whom we live, and move, and have our being," and which, if but for an instant withdrawn, the whole material universe and its forces, in all their vastness, glory, and beauty, would collapse and sink in a moment into their original nothingness.

In the meantime, I think the best way for us is to adopt the luminous definition of the man of genius, J. R. Mayer, viz., "Force is that which is expended in the production of motion; and this which is expended is, as cause of the effect, equal to the motion produced."* The idea of expenditure is thus the cardinal point in all exercise of force, and if this is constantly kept before our eyes, we shall escape the difficulties which are apt to entangle us in the distinction of property from force. The majority now hold that that which is expended is simply motion, and would describe the forces as motions. The mathematicians prefer to apply the word force to the dead pull or pressure which is the antecedent of motion. For us it is, however, far more convenient for the understanding of the phenomena, to keep to the idea of expenditure and of motion; but the grand difficulty is how to conceive of the passive forces as motion, i.e., those in which the movements of masses or atoms of matter cannot possibly produce the result, e.g., the attractive forces of gravitation, cohesion, chemical attraction, &c. For the discovery of the mechanical equivalent of heat has given the death-blow to the notion of all inherent powers of attraction and repulsion as properties of matter as well as all action at a distance, which were till lately generally held, in spite of the strong objections of Newton himself. It is not necessary we should have any theory of the cause of gravity, nor of the nature of force, in order to accept Mayer's definition, which is, indeed, framed to avoid all speculations, but it clears our conceptions to have some hypothesis.

* " Mechanik der Wärme," p. 265.

The dynamic theory of gaseous elasticity, diffusion, and
pressure opens the way for a general conception of the
manner in which all the forces may depend on movements of the matter or the æther, and I may, therefore,
briefly refer to Professor Challis's hydro-dynamic
theory* without presuming to give an opinion upon its
absolute truth, for even as a mere diagram it would
illustrate the subject better than a train of reasoning.

The hypotheses of his theory are these only : "1st, All visible
and tangible substances consist of inert spherical atoms of constant form and magnitude (for each kind) ; 2nd, All physical
force is either active pressure of the æther, supposed to be a
continuous elastic fluid having the property of pressing always
proportionally to its density, or passive resistance of the atoms
to such pressure due to their inertia and constancy of form"
(104). The existence and qualities of the atoms must be taken
as ultimate facts, and out of the province of à priori investigation (105). Professor Challis thus uses the word force in the
sense of the antecedent pressure which produces motion, while,
as above said, others think it more convenient to apply it to
the motion itself (Grove), and Mayer to "what is expended."
But, practically, it comes to the same thing, for all the phenomena by which we recognize force are traced by Challis to
movements of the æther or the atoms. Thus heat and light
are movements of the matter and æther of a vibratory kind,
while electricity, galvanism, and magnetism are similar but of
a more complicated character. In these Challis does not differ
substantially from other authors. But with respect to the
passive forces, viz., the molecular attractions and repulsions
which it is impossible to conceive as produced by movements
of the atoms themselves, and which have been hitherto the
grand stumbling blocks to the full comprehension of the nature
of force as a whole, he holds that they are produced by undula-

* "Essay on the Mathematical Principles of Physics" (Cambridge,
1873).

tory movements of the æther of a *translative* character. This he supports by mathematical reasonings which cannot be entered on, but the gist of them is that the atoms are driven together or asunder by waves, which, as it were, beat more on one side than the other, and thus push and keep them together or asunder by this perpetual impact. In this manner we may imagine the attractive forces of chemical attraction, cohesion, and gravitation, and the repulsive force of heat to be produced by movements of something, though not of the atoms which are here passive. The common statement that when a weight is raised the force or energy required to raise it becomes potential, and is again transformed into actual energy when the weight is let go and it falls to the earth, is a mere statement of the fact, and so far from being an explanation, it almost universally induces the belief that some attractive power is accumulated in the weight itself or the earth which pulls it back. Whereas by the above hypotheses of translative æthereal waves, the body is merely placed in a position in which, having distance to traverse, it can take on as actual movement in space, the motions constituting the force of a given quantity of the æthereal gravitation waves, which are thus extinguished (like the luminous waves in the dark lines of the solar spectrum) in the process of pushing the falling body and the earth together. In this manner, whether the theory be true or false, we can form an idea how something of the nature of motion can be expended in the apparent attraction of gravitation. The above two fundamental hypotheses may be disputed, and are disputed, and people are not content to accept them as beyond the province of speculation at all events. But there are no alternative speculations on the æther which do not leave quite as much to be postulated, and if we object to the hard, round, spherical atoms, and prefer the idea of permanent vortex motions in a frictionless fluid, we can suppose these acted on by translative waves in the same way.

In visiting Manchester, long before seeing a similar remark by Dr. Carpenter, I have often been struck by

14

the singular advertisement to be seen here and there : "To let, a room with POWER." Power in what sense ? Not surely the ability to spin or weave, or turn or grind, or stamp or electroplate ? Certainly none of these things, but only the force whereby any person with the mental capacity and the materials needful may do one or all of these things. Here, then, we have, valued against a given weight of gold, force in its simplest aspect—that is, motion measured as $m\,v^2$ or mass into the square of the velocity—a wheel or shaft of a certain size turning at a certain velocity. In purchasing force, therefore, we purchase no power of directing it towards any particular object. To do that, we find we have no power except by taking advantage of the pre-existing properties of matter which give it the capacity of taking on motion in a particular way after its kind. In fact, the properties of different kinds of matter may be defined as the capacity for being moved each in its particular way ; while force may be defined as the motion of matter, including the æther ; and it thus may be molar, molecular, or æthereal. The indestructibility of force is thus a corollary of the inertia of matter, for if matter cannot start into movement of itself, neither can it stop of itself, but only if its motion is transferred to other matter or æther. The conception of a mass of matter in motion from place to place, and the transference of its motion to another mass by impact, involve no mental difficulty requiring discussion. Nor does directing the motion by taking advantage of the physical properties of rigidity, &c., of matter in the form of machinery or tools. At the same time, the impossibility of attributing anything directive or formative to mere force in all mechanical construction is so plain, that to talk of force turning a chess-man, stamping a die, weaving a web, &c., by itself, would be simply ridiculous. If we wish to convert our purchased velocity into heat, or electricity, or magnetism, it is still plain that we can only do so through the intervention of the pre-existing properties of matter, and in the last we are tied down to narrower limits of the means. In dealing with molecular movements, or forces, we find that although molecules will certainly not move without force, yet they will only move in the way for which they are fitted to move by their inherent properties or qualities, no matter how the force is furnished. A fluorescent body exposed to the non-luminous, highly refrangible rays of the spectrum, will take on the motion, certainly, but not at the same rate of vibration, and thus give them out as visible light; and we have no power by any manipulation of force to cause the said body to act differently. In fact, neither can it help itself; it is just as passive in its power of directing motion. It is only when we come to chemical action that any difficulty or ambiguity has arisen, and this is simply from the loose way in which the word force has been used to denote property as well as force, and if we carefully distinguish between these, there is no real difficulty in the question. Unfortunately, even deservedly high authorities have fostered error by inadvertence in this respect. Faraday, for example, speaks of the conservation of chemical *force*, and instances the unchangeableness of the inherent qualities, which specifically distinguish oxygen from other substances, as a proof of conservation of *force*. It is needless to say the word is here used in the sense of property, but the effect is unfortunate, and has quite misled Bence Jones, who, in spite of the views of Newton, that gravity cannot belong to the essential, inherent qualities or properties of matter, because it is capable of diminution, classes weight with chemical "force" (property), saying that the union between matter and gravity is just as inseparable as the union between matter and chemical "force" (property), and that matter without weight is not matter at all. This is erroneous, and misleading in every particular. Matter without weight is not only conceivable, but it is easy to calculate the position in space where matter can have no appreciable weight at all. And it is perfectly easy to conceive, although no such state exists as a reality, matter totally devoid of all force, and in which it would retain all its inherent qualities, or properties, unimpaired. The conception of matter emptied of all force is not more difficult than that of the absolute zero of heat, which is a postulate in physical science. The truth is, there is no such thing as chemical force at all as an inherent property, any more than weight as an inherent property of matter. What is meant is, that matter has an inherent capacity for being moved by the respective forces indicated by these terms. In the case of gravitation, the particles of all kinds of matter have the same capacity for being pushed towards each other by the force of gravitation, as illustrated in Professor Challis's theory. Whereas, in chemical attraction, it is in virtue of a specific property in each kind of matter (possibly mere

difference in size) that certain kinds have the capacity
for being pushed towards each other by the force of
chemical attraction.   As I have said elsewhere,*—

" Here also two distinct factors are involved, viz., the property
of chemical affinity on the one hand, and the force of chemical
attraction on the other, the product of these two being chemical combination.   The act of chemical combination implies a
certain amount of motion, which, again, involves the expenditure of *something*, and this something cannot be the inherent
property which constitutes the chemical relationship or affinity
between the combining bodies, because that property remains
undiminished and unchangeable.   What is expended must,
therefore, here, as elsewhere, be *force*.   The inherent property
in this case cannot, any more than in that of gravity, be in itself a power of attraction, but only determines the effect of the
acting force.   This force is the same for all chemical combinations, and is therefore called the force of chemical attraction,
just as we speak of the force of gravitation, but it is the property of affinity which determines what particular bodies shall
combine with one another, using in the act of combination the
requisite amount of the common stock of force.   No amount
of the force of chemical attraction will cause the combination
of bodies destitute of affinity for each other.   To speak of the
force of chemical affinity already implies an hypothesis which
the foregoing considerations show to be erroneous.   In strictly
accurate language, we should say every act of chemical combination involves the existence of the property of chemical
affinity, and the expenditure of the force of chemical attraction.   What the nature of this force of so-called attraction
may be, is not decided, but it probably is caused, like gravitation, by pressure from without upon the particles of matter "
(p. 99). . . . . " When we thus separate distinctly the
ideas of property and force, and confine the latter to its strict
meaning in physics, we perceive the completely subordinate,
although still essential, part it plays.   Force can have no power

* " Life, and the Equivalence of Force," p. 98 and p. 189.

in producing, or developing, or altering the character of chemical
affinity in itself.   Nor has it any influence (unless for destruction) in chemical processes, except in furnishing 'that which
is expended in the production of motion,' and thus giving the
means of the display of the properties of the elements. Force
is the mainspring of the cosmical machine, but not the works.
It is in the properties of matter that lie the wonder and mystery
of the universe" (p. 189).

The cardinal point is thus to keep in mind the
absolute distinction between force and property.   The
active forces with which we are familiar are few, are
transferable, transformable, and commensurable in all
forms, and, in the form of heat, capable of dissipation
as far as our world is concerned; but as they are all
merely affections of matter, including the æther, it is
of course impossible they can exist *per se*. On the
other hand, the properties of matter, as insisted upon
by Newton, are peculiar, inherent, incapable of increase
or diminution, or of dissipation, and incommunicable:
matter is, therefore, perfectly conceivable as existing
without force, but the state of the universe would
then be chaos.   There seems no room for confusion of
the two, so far, but there are also an infinity of
secondary, or acquired, states of matter, which are also
called *properties*, and it is in reference to these that
constant vigilance is needful, otherwise we fall into the
error induced by calling two different things by the
same name.   These secondary properties are all developed by the union of force with matter, and cannot
exist without force, whereas, as before said, the ultimate inherent properties are essential, and quite conceivable as void of all force.   Among the secondary

properties, solidity and fluidity are easily seen to depend on a balance of the attractive force of cohesion and the repulsive force of heat, and therefore it is plain that the difference of inherent properties of bodies in these respects is simply *how much* of the force of heat the particles of each body require to take on before their cohesive force is so far overcome as to make their state of aggregation fluid. It is more difficult to analyze the so-called property of weight, and perceive that it is made up of the capacity of the ultimate particles to be driven towards each other by the force of gravitation, and of the presence of that force. But it is in the development of new properties by means of the force of chemical attraction that the most complete examples of the properties acquired by the union of force and matter are displayed. Every combination of different elements gives rise to different relations of the compound to the active and passive forces, and to their affinities to other bodies, and this leads up to the vast variety and complexity of the compound bodies we already know ; while no knowledge of the chemical powers and properties of matter we already possess enables us to place a limit, beyond which new powers may not be developed by new combinations and conditions. Here, however, it is always the inherent property or capacity of the substance to be affected by the force of chemical attraction, viz., affinity, which, as above said, is the determining cause, and, no matter how complex the compound, that must always depend on the property of the ultimate atoms of which it is composed. The potentiality, therefore, of all possible powers of compounds to be developed by the union of the property of chemical affinity and the force of chemical attraction must always be traced back to the inherent properties of the ultimate elements. Therefore, although force is essential to all the physical and chemical states of aggregation and combination known to us, it is incapable of being the directive or formative cause of any of them. The spherical form of small masses of fluid is not caused by any "force of sphericity" forming them into that shape, but simply results from the adhesive attraction of the molecules overbalancing their weight; and the shape of a crystal is not from any formative power of "a force of crystallization," but simply from the attractions and polarities of the molecules depending on their specific properties coming into play on a given diminution of the repulsion of heat. And so on through the world of variety of chemical compounds, in no case is mere force, or "that which is expended in the production of motion," the determining cause of any form, shape, or specific affinity.*   I may conclude

---

* The foregoing views on the necessity of the radical distinction of property and force being always kept in view, and the impossibility of any directive power in mere force, were published in 1871. It is only in biology that difficulties are apt to occur, and render such observations necessary; accordingly the subject is seldom touched upon by the mathematical physicists, but in July, 1872, Mr. James Croll wrote a paper in the "Philosophical Magazine," entitled "What Determines Molecular Motion?" which has great interest for us here as bearing on the subject. Mr. Croll has since stated that he published similar views fifteen years ago, in his book on "Theism," which I have not seen: it is now out of print. In answer to the above question, Mr. Croll enlarges on the prevalent error of attributing everything to force, and insists that far more depends on the *determination* of force than upon its *existence;* "and therefore, unless force be determined by force, the most important element in physical causation is a something different from force." This he reasons out, and concludes that "the action of a force cannot be *determined* by a force, nor can motion be determined

by illustrating, with the diagram formerly given, the above views of the subordinate nature of force in the development of the secondary properties of matter, and its dependence on the determining powers of the inherent properties of matter in all cases :—

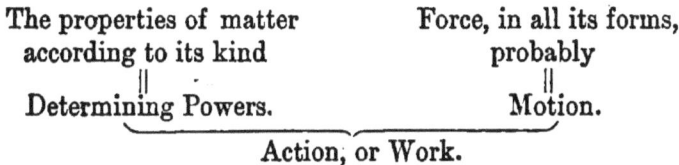

The properties of matter according to its kind	Force, in all its forms, probably
‖	‖
Determining Powers.	Motion.

Action, or Work.

The application of the foregoing to our subject is obvious, and it is unnecessary to go into detail in showing that the expressions of physicists who speak of force derived from the sun, or elsewhere, forming and building organic or any other structures, are merely figurative, and that all they really mean is that the quantity of "that which is expended in the production of motion," which is essential to the existence of all material compounds, protoplasm or other, is thence derived. In a passage to this effect, Tyndall says plainly that "the energy is conditioned by its *atomic machinery,* so as to result in the formation of," &c. This is entirely overlooked by Beale in commenting on the said passage, as if force *per se* was the directing agency. I have no doubt that when Tyndall comes to study the question for himself, and not to

---

by motion :" nor can there exist any "structural" or "formative" forces, nor "crystal-building" force, &c. ; nor can force be self-directing. In the formation of a crystal, all that the forces can do is simply to pull ; but it is the specific property of the constituent particles which directs the forces how and where to pull. "The production of form and arrangement of parts by a force is what never is, or was, or can be effected."

take his physiology from text-books, which echo the voice of the party which may be in fashion for the day, he will then see that no atomic machinery could possibly so condition the forces as to grow and act like living matter, except an atomic combination, utterly distinct from all ordinary chemical compounds, such as that now called protoplasm. He will then understand Beale's merits and position better, and, at the same time, I trust Beale will be brought to see that his "vital power or force," which is certainly not a force, and which he himself denies to be an immaterial essence or principle, superadded to protoplasm, is as superfluous as it is unintelligible.

Dr. Beale through all his numerous writings speaks of life as a force or power, and in his third edition of "Protoplasm," 1874, he thus states the question between the vitalists and non-vitalists:—" Let me first state broadly the two antagonistic and incompatible doctrines concerning the nature of everything that is alive. The one which is undoubtedly just now the most popular, is, that living matter and non-living matter alike consist of the ordinary matter and forces of our earth, and that the living and the non-living should be included in the same category. The other is, that in things living, in addition to inorganic matter and inorganic forces, is what may be termed *vital force* or *power*, which, unlike any ordinary force, is separable from the matter with which it is temporarily associated, and therefore is in its nature essentially different from every form, or mode, or mood of ordinary inorganic force" (p. 17). In other places he terms it a *"separable living force;"* and

again, " a power, force, or property of a special kind ;"
and he protests against the ridicule with which, he
says, these opinions are received. I am one of the last
who would be disposed to ridicule any opinion coming
from such a source, but I think comment is allowable.
In the first place, the words life, force, and power are
used in such a variety of senses, that, to analyze
them all, would take too long, and lead to repetition,
so I may refer to the former part of this chapter, where
life is defined as an action, and therefore can be neither
a force nor a power: force, again, is, as above said,
always separable from matter in idea, and active forces
are so in reality, and Beale himself constantly argues
against the physico-chemical school that life cannot be
a force in the only proper physical sense of the word.
So he simply uses the word force, and also property,
as synonymous with power. Now the word power is
in common life used in a triple sense, meaning at one
time the ability to do a thing if the force were
furnished ; at another, simply the force necessary for
the work ; in a third, both together. In the sense of
force it has already been illustrated. Dr. Beale uses
it ambiguously, but often in the last sense, for he
attributes to living matter a power of spontaneous
movement, while at the same time he denies that it
can create force, and is thus inconsistent. But at any
rate it is always with him a something separable from
matter, yet not a force ; which can cause matter with
force to grow, develop, and form structures simultaneously in all their parts, which shall be harmonious
when complete, and parts of which structures shall be
of no use for perhaps years after they are formed, and

then spontaneously take on in perfection their pre-ordained functions. Such a power, if separable from matter, must be an entity, and not only one with power, but intelligence—a demiurgos, in fact. At the same time, Beale expressly cuts himself off from the more consistent forms of the older vital principlists, for he rejects the vegetative, sensitive, and rational principles, which, being central, might be supposed more capable of watching over the harmony of the separate parts, and the unity of the individual. On the contrary, Beale's vital power inhabits only the individual separate molecules, and cannot be trans-ferred to the smallest distance, nor pass from one piece of matter to another, except by descent. He is thus in exactly the same difficulty as the physiological materialists were as to the process of building up of the higher orders of individualities and constituting species, before the theory of Darwin, for on this he looks askance. Unless, indeed, he considers the com-munication by descent a saving clause, and thus the development of all parts from the ovum by growth and subdivision gives the vital power that central har-monizing influence ; but again, if so, what is it but the old vital principle ? The supposition of such a power is no help in the explanation of any phenomena, as it stands in need of explanation just as much as they do, and we fear it is nothing but one more example of "the insuperable tendency of human mind to per-sonify its abstractions."* Its position over against the

---

* Fletcher frequently brings forward this saying, quoting from Barthez, who gives us one of the best illustrations of the truth of it, for he was the inventor of the term " vital principle," and the meaning attached to it.

theory of Fletcher will, I think, be made more plain
in reviewing, in the same sense as Dr. Beale, the
physico-chemical theories of a vital force supposed to
be correlative with the physical forces. Professor
Owen speaks of vital modes of force correlative with
physical and chemical modes of force, and conveys the
idea that otherwise inanimate matter—of the tissues,
for example—can show the phenomena of life when
under the influence of these forces, and cease to do so
when they are withdrawn, just as a piece of steel may
act as a magnet or not. This same idea, although not
so openly expressed, pervades the ordinary physico-
chemical school, who are not prepared to allow that
the proximate principles do not exist as such in the
protoplasm. But apart from the origin of these peculiar forces, it is easy to show that nothing of the nature
of a force can possibly have the power attributed to
them. For, in the case of harmonic vibration, as when
a bell or stretched wire gives out the same note that is
sung or played, the force of the sound-vibration of the
air passes to the bell or wire, which is capable of vibrating at the same rate. In this case the strength of the
voice sound is weakened by *exactly* the amount of
force transferred from the aërial vibrating particles to
the solid ones. Precisely the same happens when particular light vibrations are intercepted by harmonically
vibrating gaseous particles in the phenomenon of spectrum analysis. The same happens in the radiation of
heat, and numerous other examples. In all these there
is simply a transfer of force. But in magnetization of
iron by blows, or the development of heat or electricity
by friction, or in fluorization, there is also transformation

of force, still, however, within the strict condition of equivalence. In all these instances there is no change of composition of the molecules—no chemical change. The particles of the wire or bell, or magnet, or fluorescent body, are merely set in motion, and when the force is withdrawn, they return unchanged to their former state, and will serve the same purpose again and again according to their inherent inexhaustible properties.

Quite otherwise is it with the vital acts. No tissue or organ can take on any vital action and cease to manifest it as influenced by any *equivalent* transformation of force ; muscular contraction, apparently so conditioned, is already excepted from vital action. In all vital action force acts merely as a stimulus not at all in equivalence to the result, and that result is determined solely by the pre-existing properties of the organic compound—properties not possessed by albumen, fibrin, protein, or any chemical proximate principle. A change of molecular composition also takes place in every vital act, as above frequently insisted on, and one that is wholly beyond the capacity of any force to cause non-living matter to undergo. So the idea that any possible form or mode of what we know as force compelling albumen to take on the utterly peculiar metabolic changes of the living matter is as impossible as, and more absurd than, the notion of turning lead into gold, by heating, or stamping, or by any force or movement whatever. Even supposing it possible that any *peculiar* force called vital could animate albumen, the ordinary school who believe albumen to exist in living matter, forget to tell us how this force originates out of ordinary force. But it is needless to dilate on the inconsistencies of an hypothesis which is never really expressed in intelligible terms, and from what has been said of the want of directive or formative powers in all force, Dr. Beale's arguments against the school who uphold protein, animated by a peculiar force correlative with heat and motion, as constituting living matter, are in reality superfluous. It is more to the purpose to notice that those who assume to be representatives of the material theory of life, and ridicule Dr. Beale's separable vital powers, do really themselves revive the old vital principle under a new name when they speak of a formative or constructive force residing in living beings. Or if that is not their meaning, they express themselves in a manner certain to mislead. Of this I have already* given examples taken from Dr. Carpenter's " Correlation of the Physical and Vital Forces,"† where he speaks of "organizing force," "constructive force," and "various forms of vital force," which are said to be derived from external heat and from food, all which terms and expressions are quite inadmissible. Sir James Paget‡ also speaks of a "formative or vital force by which the energy in food is directed in transforming the matter of food." This force in the adult is, he says, the same "which actuated the formation of the original tissues in the development of the germ and of the embryo " (46). Again, he says that this "vital or formative force is in constant operation," and "under its direction," forms and dimensions are assumed which depend also on the

* "Life and Equivalence of Force," p. 190.
† "Quarterly Journal of Science," 1864.
‡ "Lectures on Surgical Pathology," 3rd edit., 1870.

composition of the pabulum, &c. We might, perhaps, suppose that the word "force" was here used, like the *vis insita* or *nisus formativus* of the older authors, to express a power depending on the properties of the organized matters, but, unfortunately, just at this point is given a note with the whole citations of the doctrine of the conservation oi force from Mayer to Bence Jones. It is therefore impossible but that the student should be misled into supposing that something existed correlative with physical force and called a force, but credited with powers far beyond the capability of any force and hitherto only ascribed to imaginary entities like the vital principle. The most singular application of the idea of vital force is that of Dr. Maudsley, who says :*—

"As there are different kinds of matter so there are different modes of force in the universe ; and as we rise from the common physical matter in which physical laws hold sway up to chemical matter and chemical forces, and from chemical matter again up to living matter and its modes of force, so do we rise in the scale of life from the lowest kind of living matter with *its corresponding force or energy*, through different kinds of histological elements with their corresponding *energies* or functions, up to the highest kind of living matter and corresponding mode of force with which we are acquainted, viz., nerve element and nerve force" (68). In illustrating this "upward transformation of matter and correlative metamorphosis of force," he says, "As one equivalent of chemical force corresponds to several equivalents of inferior force, and one equivalent of vital force to several equivalents of chemical force," &c. And again, "A single monad of the higher tissues would equal several monads of the lower kinds of tissue or several equivalents of its force."

From what has been said, it is obvious I must totally

.* "Physiology and Pathology of Mind," 2nd edit.

dissent from the whole of this, and to give reasons in full would be simple repetition. It would be interesting to hear an example of the chemical force which is equal to several equivalents of "inferior" force, an expression which I cannot understand. Beyond that the word "force" is simply used instead of property, and the expression equivalent is out of place; otherwise, if the potential energy of the more highly organized forms of protoplasm were really greater in proportion there would be no difficulty in finding it out and measuring it in heat units at death. Imagine the tremendous energy shut up in the brain cells of a man, to be discharged at the moment of death! why his skull would surely be blown to pieces! Seriously, there is no reason to suppose that the more complex and highly organized protoplasm contains more potential energy than some of the tissues and secretions. On the contrary, it is probable that the hæmoglobin and some other quaternary compounds are of higher potential energy, and it is certain that fat is, and the uncombined elements are higher than that. It is certainly not impossible that there may be some organic compound of higher potential energy than the uncombined elements, for such occurs in some chemical compounds (e.g., carburet of sulphur and protoxide of nitrogen); but none is known as far as I am aware. To conclude this part of the subject; the expression vital force is false and misleading in every way, and in respect to the characteristic powers of living matter, the law of conservation and equivalence of energy has no more direct bearing than in analytical chemistry. In the constitution of the protoplasm force is, indeed,

15

necessary, but not more so than to the simplest chemical compound. Force is to life as the organ-blower to the musician—essential, indeed, but utterly subordinate.

There remain still a few words to be said on the really slight difference which divides the protoplasmic theory from that of Beale, but though slight at the origin, yet just as two streams close to each other at the watershed, soon diverge widely, the difference leads to a total estrangement in their views of general philosophy. Beale and Fletcher both believe that the protoplasm is the sole seat of vitality, and both believe that the ultimate elements composing it are not chemically combined into the proximate principles we find therein after death. With Fletcher this difference of composition is simply taken as the cause of the different properties of dead and living matter. But with Beale that is not allowed to be sufficient, and a " separable force or power " is postulated. This, he admits, cannot be a force, nor does he allow it to be a result of the properties of protoplasm, therefore it must be an entity. When new living matter is formed from pabulum, a portion of this separable force or power is supposed to be transmitted from the pre-existing living matter which has the faculty of creating an unlimited quantity of it out of nothing, while at death it is again annihilated, contrary to what obtains of ordinary force, which may change its form, but cannot cease or be annihilated (" Microscope," p. 324). Some forms of vital power, however, are immortal, viz., that in the gray matter of man's brain " may be freed from the material and yet exist without cessation, extinction, or annihilation." Putting aside for the present this reference to the soul of man which he has no warrant for placing in the same category as mere life, which latter we possess in common with animals and plants and which is certainly mortal, this revival of a spiritual principle or essence of whose nature we can form no conception and which explains nothing, is a very strange thing in the present day. If we build a house we do not require to add a separable force or power of edificiality, nor if we make any chemical compound is any such thing additional necessary to give it its special qualities. The living actions, says Dr. Beale, cannot be explained by the properties of the matter of the protoplasm, these are permanent endowments, while the vital properties are superadded temporarily, and when removed cannot be restored. The properties of sugar or quinine may just as well be said to depend on the addition of a power of saccharinity, or of bitterness and fluorescence, for these properties are no more permanent than those of protoplasm. The moment these bodies are decomposed, the said properties are gone. But, says Dr. Beale, you can make these bodies again, and in that case the said properties would be there again, and thus they may be said to be permanent, whereas you never can make protoplasm again when once destroyed. As a matter of fact, we cannot make sugar or quinine again when once destroyed, and may never be able to do so, and we may grant that we never shall make protoplasm, but that does not alter the reasoning that if these compounds are by any means put together, they would manifest their specific properties, the one dead and the other living. There is no break in the reasoning. The truth is, that all Dr. Beale's reasonings.

on this point are simply variations of one single argument, viz., the power of self-renewal from heterogeneous material and of germinal development, are so contrary to all the ordinary chemical powers of matter, that it is impossible to conceive that any possible complexity of molecular constitution, or state of affinity such as here called metabolic, could produce such effects without the addition of a superadded immaterial entity. It would be easy to quote pages in support of the statement that this is his position, but it would be superfluous. The question is not susceptible of proof by argument, and the material theory rests simply on the fact that vitality is never found without protoplasm, and the latter never exists in its integrity without vitality, and yields nothing but ordinary matter on its destruction. If you assume a spiritual principle in addition, whose existence is beyond proof to the senses, you are bound to show why the same should not be assumed as the cause of the special qualities of ordinary chemical compounds. The difference between Fletcher and Beale is entirely on a hypothetical point, which can never be brought to any issue or submitted to the test of experiment, whereas on all practical subjects they are agreed—both *vitalists* in the sense that true living matter is a substance existing in a state quite different from that of ordinary chemical combination, reacting with ordinary matter and force in a way entirely *sui generis* and incapable of entering into any chemical combination without destruction. The summary of the difference between Fletcher and the physiological materialists on the one hand, and Beale on the other, is simply that the one party profess they do not know to what extent the properties of matter may be developed—even to the extent of producing the phenomena of life—while the other, Beale and the vital-principlists, profess to know with certainty that a new principle is required. Under these circumstances, we read with surprise the conclusion of Beale's chapter vii. of 3d edition of "Protoplasm." "Any one who will contemplate such an arrangement of tissues as that which may be demonstrated in a specimen like the one figured [tongue of Hyla], will not rest satisfied with attributing it to the 'properties' of the elements entering into the chemical composition of the substances out of which it has been made. The 'property.' hypothesis accounts for absolutely nothing. Its advocates are unable to explain how one of the tissues has grown into the form it ultimately takes, how it acquired its structure, or how it came into relation with adjacent textures. No wonder the disciples of 'property' philosophy pride themselves upon the interest they take in the broad general features, and try to make the public believe that they have reason to look down upon minute details and contemptuously disregard the facts demonstrated by those who study the structure of the bodies of living beings.

'The simple fool is he who knows that he does not know,
The compound fool is he who does *not* know that he does not know.'"

I apprehend that Fletcher belongs not to the second category, though he need not be ashamed to be classed with the many wise, and good, and great who confess to the first.

Relation of Consciousness to Force.—"Thought," says Fletcher,* "is not, any more than life, anything substantial, but an abstract term by which are signified certain phenomena peculiar to the higher orders of living beings, and necessarily resulting from one property of their organism, viz., the faculty of thinking in action" (p. 92). "The alleged proof advanced by Malebranche that ideas must be substantial, since they have distinct properties, is quite untenable. Ideas have no properties. They are nothing but a mode of existence of the thinking organ, acted upon by its proper stimulus, and no more substantial than combustion or motion, which are, in like manner, modes of existence of the thing burning or the thing in motion" (p. 113). "We know matter, it is said, only by its properties, extension, impenetrability, &c.; and we know mind also only by its properties, reason and passion; and where the two sets of properties are so decidedly dissimilar, they must indicate, it is argued, different entities. Thought, therefore, may be *attached to* matter, but it cannot be a *mode of being* of matter, since matter in no case betrays those indications by which we recognize mind. In *no other* case, certainly, for in no other case is the organization of matter such as to be susceptible of this mode of being"† (p. 92). Such are the opinions of Fletcher on the much-vexed

---

* "Rudiments of Physiology," Part III.

† "If thought is to be called a function of matter, it must be acknowledged to be a function wholly peculiar and unlike any other" ("As Regards Protoplasm," 2nd edit., p. 46). Well, why not? The spiritualists tacitly assume that merely to state the proposition is to prove thought to be the function of an inconceivable something called spirit, of which they, as men of science, know nothing, and can know nothing.

question of the relation of mind to matter, and I apprehend they are in harmony with what is expressed, but not better expressed, by the majority of philosophical thinkers of this day. I do not propose to enter on the vast and intricate question of the nature of mind, but merely to comment on the relation of one of its phenomena, viz., Consciousness, to force. We may pass by a large number of truly mental phenomena as not being necessarily attended with consciousness, and in these the relation of the transformation of Energy, or the doing of work, is probably the same as obtains in ordinary vital or metabolic action, viz., the ingo of force through stimuli and pabulum exactly balances the outgo in the form of heat, mechanical movement, and the potential energy still remaining in the living matter, or its products, while nothing is counted for the peculiar properties given by the state of organism. All the discharges of the reflex and commissural stimuli which are essential to bring into play the numerous elements of the very simplest act of thought, whether conscious or not, and the forcing the atoms into new positions, implied probably by memory, and all expression of thought and emotion, whether by voluntary or involuntary movement, or by sympathetic stimulation, have to do with work obviously enough. But it is with consciousness that the true difficulty lies. Is it allowable to suppose that with each act of consciousness there is a consumption of a given amount of energy and reappearance of it again, as heat, or other mode of force, on the cessation of consciousness? Can we admit the propriety of Mr. Huxley's expression, " the mechanical equivalent of consciousness ?" I

apprehend not ; and it seems to me that the parallel
mode of expression would be the "mechanical equivalent of fire," not of heat.   The pursuit of this analogy
may lead us up to the point where mind and energy
can be seen to diverge.   To the ancients, with whom
fire was an element, the analogy between it and life
was always deemed close, and their ideas survive even
to this day in the popular and poetical terms applied
to life.   But with the rise of modern science the
notions of a special substance, called fire, or of any
occult qualities connected with combustion, have been
dissipated.   The phenomena are presumed to be accounted for when we describe the new compounds
formed during the chemical combination of the combustible and oxygen, and measure the active force
evolved as heat and light.   Nothing is counted for any
perception by the elements themselves of the mutual
influence of their affinities.   There is nothing to correspond to fire.   It has no existence except as a word
describing phenomena whose causes and consequents
are otherwise sufficiently accounted for.   Combustion
may stand for the type of all strictly chemical and
physical processes in the world : in none of them, as
far as we know, is there any phenomenon like perception by themselves of the action taking place
between the different elements.   When we advance to
the infinitely more complicated molecular state of
living matter, we still find in the whole functions of
organic life the same relations of matter and force.
The involutions of the self-renewing protoplasm, and
its subdivision, growth, and development, the deposition
of tissue and secretion, and all the functions in animals

depending on involuntary motion and organic sympathy, fully account for all the matter and force, entering in as pabulum, stimuli, and conditions, in the form of results (granting, of course, that the mere fashion of molecular arrangement can account for properties so different from common chemical ones), without any consciousness of perception. Even when we rise to the reflex actions closely connected with sensation, the central nerve cells of the spinal cord may be so organized that harmonized groups of muscular movements may be performed on the application of a stimulus without any consciousness. And beyond that even, as above said, certain processes of thought may take place without consciousness. In all these cases the transformation of force as work done in discharge of nerve force, in nutrition and secretion, in growth of new protoplasm, and in forcing atoms into new positions in the molecular organization of existing protoplasm, is an essential and integral link in every process. But now we come upon a totally new phenomenon. In the most highly-organized matter, viz., the gray matter of the sensorium, not only do the stimuli cause the same metabolic, or vito-chemical molecular changes as in other nerve protoplasm, but now, for the first time in the long history of the development of new properties by the complexity of molecular constitution, we meet with that of conscious perception of those very molecular reactions. It is, therefore, not a link in the chain of material transformations, but an *incidental* phenomenon manifested in certain changes, whose causes and results are otherwise fully accounted for. It seems to be a property only perceptibly manifested when a certain degree of activity of metabolic change is going on in its peculiar kind of nerve protoplasm, which, like all other, is liable to a *plus* and *minus* of degrees of activity, according to the degree of stimulation, and in proportion to these it appears susceptible of exhaustion and fatigue from action and regeneration during rest. The phenomenon of Consciousness may be likened to a loud sound, or a brilliant light accompanying the working of a steam-engine, but for which no deduction is to be made from work done. To carry out the analogy, however, the sound and light would need to be inaudible and invisible, except to the machine itself, and then we see that no force can be expended on it. So with consciousness: it is a purely passive phenomenon, and has no power of making its existence known except through transformation of pre-existing force residing in the protoplasm and pabulum, of whose reaction it is itself an incidental phenomenon. It is, therefore, not work, and does not consist in transformation of force, and cannot be said in any sense to have a mechanical equivalent, inasmuch as it is not convertible into any mode of force.* Consciousness thus

---

* It may be likened to the action of a mirror in reflecting the rays of light, or to a pillar supporting a statue, or to a vessel inclosing a gas at a higher pressure than without. In all these the inexhaustible properties of matter are alone in play, but if the light were absorbed, or the pillar yielded, or the vessel burst, expenditure of force would take place, and work be done. If, therefore, any disappearance of force took place during consciousness, unaccounted for by the heat, work, and products of cerebral action, and reappeared again as heat, or other force on its cessation, then we might speak of consciousness in connection with the word work. As it is, no such consumption of force is proved to take place, and therefore consciousness must belong to the properties of matter, not to force, although it can only be manifested during a certain extremely complex series of metabolic changes, for which, however, force is essential as a subordinate factor.

belongs to the properties, not the forces, of matter, if
to matter at all, and is equally out of the province of
à priori investigation with the qualities of the æther
and the atoms which must be postulated as the necessary foundations of all scientific inquiry.   Already, on
certain simple hypotheses of the æther and atoms, an
intelligible theory of the whole physical world is being
built up, and glimpses are seen of similar explanations
of chemical action.   The vital and mental modes of
being of matter are not yet intelligible on any
hypothesis of the nature of the atoms, but we have
the same grounds of experience for putting them into
the category material as we have for the chemical
functions of atoms.

If experience so teaches, we must postulate that the
Almighty created matter and force so constituted that
by their reaction they were capable of producing the
whole phenomena, not only of physics, chemistry, and
life, but even of mind; and if He pleased so to do, who
shall gainsay Him ?   It is for us to investigate and
understand His works, as far as our faculties enable
us, and not to criticize or place limits to His power.
The arguments that the brain is not merely the substratum whereon some unknown entity may rest, have
been reviewed again lately by Claude Bernard, who
concludes almost in the words of Cabanis : " The brain
is the organ of the mind in the same sense as the heart
is the organ of the circulation, or the larynx is the
organ of the voice " (385).*   At the same time, he
makes no pretension to understand how it acts in
thought and consciousness.   It is needless to recapitu-
late these arguments, as they are accessible through

* "Revue des Deux Mondes," Mars, 1872.

most works on physiology. They are strengthened, also, by facts long known, on which the scientific part of phrenology is founded, as well as by the more recent experiments showing the localization of different faculties in the brain. They are also illustrated by the observations of the eminent mathematical physicist Professor Tait, although, certainly, these were made for a different purpose. When asked "to face the question, where to draw the line between that which is physical and that which is utterly beyond physics, again our answer is, experience alone can tell us, for experience is our only possible guide." He defines the proper field of physics "as concerned solely with matter and energy and the phenomena depending upon these;" and adds, "All our reasonings in physics *must*, so far as we know, be based upon the assumption, founded on experience, that in the universe, whatever be the epoch or the locality, under exactly similar circumstances exactly similar results will be obtained" ("Brit. Assoc.," 1872, p. 7). I submit that experience shows this to obtain of the animated world also, for throughout the whole of it, both plant and animal, a similar structure and composition produce exactly similar results, both bodily and mental; except as regards the immortal soul of man, and there all knowledge from experience is wanting. I will go farther, and say, that if a Frankenstein could construct some animal—say a dog or an elephant—particle for particle, exactly the material counterpart of some existing one, it would have exactly the same, not only bodily appearance and powers, but also mental powers, including memory and acquired habits. All experience teaches us that countless myriads of generations of the lower animals, made up of certain complex material compounds, live, and move, and display conscious thought in their day, and then as individuals, perish utterly, like a rain-drop, a crystal, or a salt, whose elements are dispersed; while their specific powers depend as much on the matter they are composed of as these simpler powers of a stone or an acid. Who shall gainsay this ?   Not, certainly, the physical philosopher, who has no experience except with matter, and knows nothing, and can know nothing, from experience of the existence of spirits.   When, therefore, Professor Tait applies the term "pernicious nonsense of the materialist" to the deduction that volition and consciousness depend on modes of being of matter, he either commits the mistake exposed by Fletcher at the beginning of this section, viz., that because inorganic compounds of matter do not show life or mind, therefore no other mode of combination such as protoplasm can; or else he falls into the common error of confounding the immortal soul granted to man alone, with the life and mind of mere animals.   This question will be touched on more specially in the last chapter of this work.

We must not disguise the fact that the view of consciousness as an incidental and passive phenomenon, still further increases, in a physiological sense, the difficulty of comprehending freedom of the will, for how can the pleasing nature or the reverse of a feeling take effect on the next step in the chain of actions culminating in the discharge of nerve force to produce voluntary or involuntary action ?   Here Dr. Beale's

hypothesis of "vital force or power" gives us no help,
for as we have seen, it is not a force, and he affirms it
is not an entity; and, moreover, whether or not, it
cannot originate any force to express itself. For he
upholds the doctrine of the conservation of energy, and
denies that the said vital power can create force, and
declares that every expression of thought, including of
course reflex central stimulation, requires the death of
bioplasm and evolution of pre-existing energy. No
physicist, as far as I am aware, meets this except
J. Herschel, who plainly gives to the will of animals
the power of *creating* energy, although the quantity
may be very small each time. If so this is capable of
proof, for the food of an animal should yield more work
or heat when consumed by it, than when its potential
energy is used in any other way. No such proof has
been given, so no other physicist has admitted an exception to the law of conservation of energy, merely to
get over a physiological difficulty. However, although
no solution of the difficulty is yet offered by physiologists, yet we must remember we do not know why the
perception of one stimulus should be pleasing or the
reverse. That no doubt lies in the nature of the molecular change set up by it, but of the nature of that
change nothing is known except by its effects, which,
however, are equally perceptible in vital action un-
attended with consciousness. Therefore the cause of
avoidance of things causing disagreeable feelings may
lie in the vital action of which consciousness is the
passive perception, and thus no greater difficulty is
opposed to freedom of choice than is already given by
the non-existence of spontaneous action without

stimuli. Consciousness is thus here stated to be quite inexplicable by all that is otherwise known of the reactions of matter and force, and therefore referred to a new manifestation of the properties of matter in the peculiar and complex compound in which it is shown, while the potentiality of it must, of course, have existed in the elements going to form that compound.*

* Leibnitz, in his "Monadologie," traces back the germs of consciousness to the ultimate atoms of matter, which he holds to be indivisible and possessing different qualities or properties, which consist essentially in perception; and that the higher living beings are made up of myriads of monads, each with a sort of perception of its own, only of lower intensity than belonged to the central monad predominant in a group and called the soul, or mind, or spirit. If these speculations give us no explanation of the difficulty in any scientific sense, yet they show us how the greatest minds have striven to bridge over the gulf between mind and matter without the arbitrary supposition of an immaterial substance added to matter. The remarks of Fletcher on this point are as follows : "The mode of existence in which perception in its most circumscribed and simplest form, as displayed by inorganic matter, consists, is exalted in irritation to another mode of existence, in which the character of the perception is altogether changed ; one step farther and the perception becomes such as to be, not, indeed, directly acknowledged by any outward manifestations, but to be for the first time more or less distinctly recognized by the percipient itself. The creature exercising it is not yet mentally individualized ; it is not yet rendered distinctly aware of its own existence as a result of this mode of existence ; but it is approaching such a sense of individuality—such an apprehension of its own being and attributes. If it be denied that any mode of being like perception can develop consciousness, it may be reasonably asked, how do we know that ? Nay, how do we know, on the other hand, that every mode of being does not develop this condition in a degree more or less proportioned to its more or less exalted character ? We are accustomed to think of consciousness only as connected with thought, and that highest degree of consciousness which constitutes thought in ourselves under ordinary circumstances, absorbs all minor degrees, and is, as it were, identified with our collective existence. 'We think, therefore, we are.' But what right have we to assert that no other degree of consciousness can exist, or be certain that that degree of consciousness which attends sensation, or even that extreme degree of it which constitutes thought, cannot result from any mode of being of either the whole or a part of our body, because no such consciousness results from any other mode of being? How do we know this ? As individualized by the exercise of thought, we cannot appreciate any minor

This is substantially the view taken by most of the physiological materialists, although from the ways they express themselves, the relation of the phenomenon to force is ambiguous.   Thus Mr. Huxley makes the mental dependent on the physical change, calling the former psychosis, and the latter neurosis; but, as we have seen, he speaks of the mechanical equivalent of consciousness.   Fechner, Bain, and Lewes speak of them as two different sides of the same phenomenon—as of the convex and concave side of a curve.   Maudsley, we have seen, speaks of higher kinds of energy involved in central operations in a way, we have endeavoured to show, is inconsistent with the nature of physical force.   He says, also, there is a fallacy in the axiom of Cabanis, because " it is plain that the tangible results of the brain's activity, the waste matters which pass into the blood for assimilation by tissues of a lower kind, and for ultimate excretion from the body, might not less rightly be called the secretion of the brain, and be compared to the bile, than the intangible energy revealed in the mental phenomena" (p. 42).   This, taken in connection with the views Maudsley has just expressed respecting the higher kinds of energy above spoken of, leave us in doubt whether he does not look upon mind as in some way correlated with physical force, in which case the

---

degree of consciousness, either in ourselves or other beings; but some degree of this state, appreciable by the beings or part of the beings, in which it takes place, is perhaps so far from never resulting from any, that it always results from every mode of existence.   There is no real violence done, then, by the apparently abrupt introduction, as a result of a certain mode of being, even of that degree of consciousness which constitutes thought, and, still less, of that minor degree which constitutes sensation " (iii. p. 11).

fallacy can hardly be said to lie with Cabanis, whose expression is : " In order to have a just idea of thought, it is proper to consider the brain as an organ specifically adapted to produce it, in the same way as the stomach and intestines are adapted to produce digestion." There is here no such fallacy as Dr. Maudsley speaks of, and it is free from the ambiguity respecting force. In fact, it is identical with the conclusion not only of Claude Bernard, but of Dr. Maudsley himself, as expressed at page 44, thus—" Nevertheless, it must be distinctly laid down that mental action is as surely dependent on the nervous structure as the function of the liver confessedly is on the hepatic structure—that is the fundamental upon which the fabric of a mental science must rest." Professor Bain, whom we may count the highest authority, rejects the supposition of a spiritual substance between the material and the immaterial shores of mental phenomena. Indeed, this once so common notion (of an immaterial entity) appears to rest on the idea that, as we know much about ordinary matter, all of which is totally dissimilar from mind, it is difficult to imagine that as belonging in any way to matter ; while, as we know absolutely nothing about spiritual substances, it is easy to clothe them with any powers we like.* But the language of Pro-

* This is illustrated by the expressions of Mr. Ponton ("The Beginning,") who adopts substantially Dr. Beale's " vital power or force," under the name of " somewhats," or " organizers." He says it is not necessary to have two kinds of "somewhats," one for giving mental power, and the other for organizing the lower bodily parts. Then, at page 314, he says—" The circumstance that volition is exercised consciously, and organizing power unconsciously, is not enough to warrant our regarding these two as being exerted by wholly distinct essences." He had just said that it was idle to attribute mental phenomena to material elements ; but he has no difficulty in attributing various dis-

16

fessor Bain is also ambiguous, for although he says,
" We have every reason for believing that there is an

---

tinct powers to " essences." To be sure he knows a great deal about
material elements, but, alas! what does he, or any one else, know
about essences? And yet how easily he settles how much each particular essence can do! Many persons—not, of course, including the
above-named — confound immaterial with spiritual. All force and
action, every event or process, including life and mind, are, of course,
immaterial in one sense, although dependent on matter. A spirit is
supposed to be a substance capable of existence *per se*, but without the
inherent properties of matter, and is therefore spoken of as an immaterial substance. The word immaterial has here a different sense.
But of such an immaterial *substance* the human mind can form no true
conception. For I hold as irrefragable the aphorism of Aristotle,
" Nihil est in intellectu quod non fuit prius in sensu ;" especially as
rendered by Fletcher—viz., the human mind can conceive nothing, the
elements of which were not first perceived through the senses. For
instance, we can imagine golden mountains, mermaids, flying fiery
dragons, and all kinds of non-existent things; but gold, women, fish,
fire, wings, and serpents were all already known to us through the
senses. In like manner all sensation depends on changes in protoplasm
produced by stimuli, which are either the incorporation of matter itself, or they are force, which is an affection of matter. We have thus
no knowledge of the external world at all, except as matter and
force, and therefore our minds can form no conception, properly so
called, of anything except what pertains to matter and force, or their
actions. The truth of this is seen at once by the incongruity and
absurdity of all attempts to picture to ourselves the nature of spirits,
which all result in merely some fanciful combinations or omissions of
the properties and forces of matter which we are familiar with in their
gross and concrete state. Witness the notion of the likeness of a spirit
to an attenuated gas or vapour, which has descended from the childhood of the world—a very unfortunate illustration to pitch upon,
seeing that a gas is the thing we know with the most absolute certainty
never to be possessed of life, or even form. Or take a more refined idea.
" Conceive," says the author of the classical *Religio Medici*, " conceive
light invisible and that is spirit " (xxxiii.). Since the true nature of
the invisible rays of light is known, can anything show more strongly
the futility of the efforts of the human mind to form any conception,
worthy of the name, of immaterial substance? Nor can we obtain any
knowledge of the nature of powers, entities, intelligences, and personalities
existing independently of matter and force, by means of natural science, for
that is concerned solely with the invariable phenomena of matter and
force cognizable to us from sensation and experience. But it is surely
a shallow philosophy which would deny the possible existence of other
modes of being than those conceivable to human beings ; or even that
human beings might attain to a knowledge of the existence of such

unbroken material succession, side by side with all our mental processes," yet he says in the same paper, "The brain is a system of myriads of connecting threads, actuated, or *made alive*, with a current of influence

---

other modes of being by revelation or by the evidence of the senses testifying to certain facts which indicated a reversal of the otherwise invariable laws of matter and force. Such facts would be supernatural, or miraculous, but would be none the less credible because human beings would still be incapable of conceiving the nature of the beings exercising those miraculous powers. However, the belief in the spiritual world does not necessarily countenance the notion of spiritual substances taking part in the ordinary operations of life and mind in the material world any more than in those of gravitation or chemical action, for the same law of equivalence of force holds good in the organic as in the inorganic kingdom. And no spirit could have any influence on mind or life, any more than on so-called brute matter, except by expenditure of force. Now this implies the expenditure of what a spirit cannot possibly possess if force is an affection of matter. If, therefore, a spirit can have any influence whatever on material beings, either occasional and exceptional, or continual as a substantial mind, it must have the power of creating, out of nothing, a measurable, indestructible something, which we know as force. Such a power has hitherto been ascribed to the Almighty alone. Doubtless at His pleasure He could endow subordinate spiritual beings with such a power, but all exercise of it would be *to us* supernatural and miraculous, and every alleged instance of it would require the strongest proof. No conception of the human mind of a spiritual or immaterial substance is at all more congruous or respectable than the vulgar notions of those mere creations of the fancy, ghosts, fairies, imps, genii, and the like. The slightest rap of spirits imagined by the so-called spiritualists, who in our day are, no doubt, the intellectual descendants of the believers in magic and witchcraft, would be as great a miracle as any recorded in Scripture. And, in short, the smallest influence exerted by any spirit on material objects or creatures, even the slight stimulus needful to excite a spectral illusion, would be a creation of force, and therefore supernatural. Whatever may have been the case in past ages, we have no warrant for the belief that such supernatural power is now given to subordinate immaterial beings to act on the material world, or its inhabitants. Science, strictly so called, affords no evidence of the existence of immaterial substances, beings, or intelligences, and for her such exist not. On this subject, therefore, appeal to science is vain. Let us be satisfied to receive in their widest sense the words of the inspired writer, "Eye hath not seen, nor ear heard, neither have entered into the heart of man the things," not only alluded to, but as equally applicable to all things pertaining to the nature of the spiritual world.

16—2

called the nerve force ; and this nerve force is a member
of the group of correlated forces." And again, that
force maintains " nervous power," or a certain flow of
influence circulating through the nerves, which circulation of influence besides . . . " has for its distinguishing
concomitant the *Mind*." Again, " The extension of the
correlation of force to mind, if at all competent, must
be made through the nerve force, a genuine member of
the correlated group."* It is difficult to guard ourselves against the idea that these expressions mean
that mind is dependent on a mere physical vibration
of matter ; also the expression, made alive by any
force, is inadmissible. The same ambiguity is to be
found in the latest expressions of Mr. Lewes, with the
additional difficulty that we cannot tell whether the
author puts them forward as an explanation or a mere
analogy.

" The great problem of psychology, as a section of biology,
is, in pursuance of this conception, to develop all the psychical
phenomena from one fundamental process in one vital tissue.

" The tissue is the nervous : the process is a grouping of
neural units."

" A neural unit is a tremor. Several units are grouped into
a higher unity, or, neural process, which is a fusion of tremors,
as a sound is a fusion of aërial pulses ; and each process may
in turn be grouped with others, and thus, from this grouping of
groups, all the varieties emerge. What on the physiological
side is simply a neural process, is on the psychological side a
sentient process. We may liken Sentience to Combustion, and
then the neural units will stand for the oscillating molecules.
Sentience may manifest itself under the form of consciousness,
or under that of sub-consciousness—which may be compared

---

* Bain on " Correlation of Force Bearing on Mind."—" Macmillan's
Magazine," vol. xvi., p. 372.

to combustion manifesting itself in flame and in heat " (pp. 135-6).*

And again, at page 119, consciousness is defined as " a succession of neural tremors variously combining into neural groups." " Consciousness may be pictured as the mass of stationary waves formed out of the individual waves of neural tremors" (p. 150).

If he means the pattern of the waves on the water, and not their motion, this would agree with my analogy with fire ; but if the motion of the oscillation is meant, I cannot admit the truth of the analogy, for all tremors, however complicated, in which the molecules return to the same state at their cessation, are merely physical force, and, as such, have no relation to life or consciousness, except that of a stimulus to excite the metabolic changes in which life consists.

* Lewes's " Problems of Life and Mind," i.

CHAPTER XI.

THE protoplasmic theory of life was brought before the general public in this country a few years ago, by Mr. Huxley, but it was not properly appreciated or understood, for some reasons which it is incumbent on us to consider now. The anatomical unity of the living matter, as stated in Mr. Huxley's lecture at Edinburgh, in 1868, was objected to by his critic, the eminent metaphysician Dr. Stirling, as only made out by keeping out of sight essential physiological differences, for if all life is identified in protoplasm, so also you must differentiate all life in protoplasms ; and instead of all kinds of protoplasm being interchangeable, as he asserts was said by Huxley, there must be infinite difference in power, form, and substance, among different kinds of living matter. This is hardly warranted by Huxley's words, and the diversity of the several kinds of living matter is an essential part of the doctrine, not only of Fletcher and Beale, as we have seen, but also of Häckel, who says :* " In short, all the immeasurable variety in the most diverse properties of

* " Generelle Morph.," i. 277.

organic bodies perceptible to the senses, which excite and delight our perceptions, is to be traced back to the alike infinitely numerous and delicate differences in the atomic constitution of the albumen-compounds which constitute the plasma of the plastids." The word albumen here will be noticed presently; but, in the meantime, it is obvious that the whole difficulty may turn on the use of the word same or identical, without sufficiently precise definition. With Fletcher there is no ambiguity when he speaks merely of the anatomical unity and of a similarity of molecular composition distinct from all chemical compounds, while capable of infinite diversity within its own region. But in Huxley's language, owing probably to being spoken in a lecture, there are some omissions that cause an ambiguity, which has misled Dr. Stirling, and others. For example, these expressions implying the sameness of the nature of protoplasm :—

"A single physical basis of life," and, through its unity, "the whole living world," is pervaded by "a threefold unity, viz., a unity of power or faculty, a unity of form, and a unity of substantial composition." "Protoplasm simple, or nucleated, is the formal basis of life. It is the clay of the potter which, bake it, paint it as he will, remains clay, separated by artifice, and not by nature, from the commonest brick or sun-dried clod."

With such passages as these, we can hardly wonder that readers might go away with the idea that the sameness of protoplasm was meant by Huxley, although further reflection would have shown that such an idea is incompatible with the protoplasmic theory

altogether. Nevertheless, Dr. Stirling so understood
him, and also Professor Gamgee thus speaks : " Looking
upon protoplasm as a definite chemical principle, Professor Huxley argues for its identity in all plants and
animals, and speaks of living and dead protoplasm."*
Dr. Beale also is chargeable with ambiguity from an
opposite cause, and has, no doubt, helped to confuse
the mind of the public, for he lays great stress on the
small difference of chemical constitution betwen different kinds of protoplasm, while he exalts the difference of power or faculty they possess in favour of
the supposed vital principle as the cause of all vital
faculty.   Dr. Stirling himself brings forward proofs
that different varieties of protoplasm do differ in fact,
and may easily be conceived to differ even though the
same elements are present, just as isomeric bodies do.
It is very remarkable, therefore, that he does not see
that his whole tract is directed not really against the
protoplasmic theory, but against, partly, a misappre-
hension of Huxley's meaning, and partly what is really
an error on the part of Huxley, viz., the giving the
name of protoplasm, not only to different kinds of
formed material such as tissues in the living body, but
also actually to dead matter, and the going so far as
to confound pabulum with protoplasm.   After following Fletcher and Beale in speaking of the consumption
of protoplasm at each vital act, Huxley, comparing it
to the shrinking of the *Peau de Chagrin,* in Balzac's
tale, says that even as he speaks his Peau de Chagrin
has been wasting, but he will presently recruit it from
mutton-protoplasm.   He then goes on :—

* " On Force and Matter," p. 16.

" A singular inward laboratory which I possess will dissolve a certain portion of the modified protoplasm ; the solution so formed will pass into my veins, and the subtle influences to which it will then be subjected will convert the dead protoplasm into living protoplasm, and transubstantiate sheep into man." Farther on he remarks—" Hence it appears to me a matter of no great moment what plant or what ánimal I lay under con-tribution for protoplasm, and the fact speaks volumes for the general identity of that substance in all living beings.   I share this catholicity of assimilation with other animals, all of which, so far as we know, could thrive equally well on the protoplasm of any of their fellows, or of any plant ; but here the assimi-lative powers of the animal would cease.   A solution of smell-ing-salts in water, with an infinitesimal proportion of some other saline matters, contains all the elementary bodies which enter into the composition of protoplasm ; but, as I need hardly say, a hogshead of that fluid would not keep a hungry man from starving, nor would it save any animal whatever from a like fate.   An animal cannot make protoplasm, but must take it ready-made from some other animal, or some plant—the ani-mal's highest feat of constructive chemistry being to convert dead protoplasm into that living matter of life which is appro-priate to itself."

These expressions would be partially applicable on Buffon's hypothesis of organic molecules pervading all living things, and forming the different organisms by being set up in different moulds, while they are scat-tered, but not individually destroyed, at death.   But Huxley repudiates that hypothesis, and the passage stands in its apparent inconsistency.   Can we, there-fore, wonder that Mr. Thornton should write thus in the *Contemporary Review,* in 1872, in a paper entitled " Huxleyism ?"

After saying that a thing cannot exist without its properties, and instancing water, he goes on :—

ᶠ "The habits of exhibiting these phenomena in conjunction with certain other habits, make up the aquosity or wateriness of water. They are part of water's nature, and, in the absence of any one of these, water would not be its own self, and would not exist. But in no such sense whatever is the life or vitality whereby what we are accustomed to call animated are distinguished from inanimate objects, essential to the existence of the species of matter termed matter of life, or protoplasm. Take from water its aquosity, and water ceases to be water; but you may take away vitality from protoplasm, and yet leave protoplasm as much protoplasm as before. Vitality, therefore, evidently bears to protoplasm a quite different relation from that which aquosity bears to water. Protoplasm can do perfectly well without the one, but water cannot for a moment dispense with the other. Protoplasm, whether living or lifeless, is equally itself; but unaqueous water is unmitigated gibberish. But if protoplasm, although deprived of its vitality, still remains protoplasm, vitality plainly is not indispensable to protoplasm—is not, therefore, a *property* of protoplasm" (p. 671).

The truth being, that the so-called " dead protoplasm," and "ready-made protoplasm," is no protoplasm at all, but merely the chemical remains of what was once protoplasm either immediately at death, or before it was converted into tissue long anterior to the death of the animal or plant: and now it is nothing but pabulum, and there is no difference between it and the smelling-salts with the other salines, except that some living organisms can assimilate the latter, while others cannot, owing no doubt to difference in the molecular constitution of their respective protoplasms.

The expressions referring life to chemical changes have also been found ambiguous, such as—" If the properties of water may be properly said to result from the nature and disposition of its component molecules, I can find no intelligible ground for refusing to say that the properties of protoplasm result from the nature and disposition of its molecules. When hydrogen and oxygen are mixed in certain proportions, and an electric spark is passed through them, they disappear, and a quantity of water, equal in weight to the sum of their weights, appears in their place. Is the case in any way changed when carbonic acid and ammonia disappear, and in their place, under the influence of pre-existing protoplasm, an equivalent weight of the matter of life makes its appearance ?"

These sentences accord with Fletcher's view, and, as far as they go, represent it properly ; but Dr. Stirling and Dr. Beale object that this necessarily implies that the process of making protoplasm is within the compass of ordinary chemical manipulation, and also call attention to the fact that water may be formed from its elements in many ways without the intervention of pre-existing water, whereas no particle of living matter can be made except by, and through, already existing living matter which grows at the expense of the pabulum, instead of being destroyed and forming a third new substance with it, as in all chemical actions. This last is quite true, and constitutes the absolute distinction between protoplasm and all other material compounds, but does not prove that it is not simply a material compound. And in respect to the first part of the above sentence, it is simply not true, for there are many existing chemical compounds which we cannot make, far less one so far out of all ordinary chemical nature. It is not necessary for the material theory that there must exist a natural process whereby the first examples of living matter were formed. There may exist such a process, or there may not; as yet none such is known to exist, or even to have existed. Fletcher is content to receive the doctrine of the inspired record, that it was fashioned from the "dust of the ground," by a power just as miraculous as would be involved in the creation of a spiritual substance called life. It is thus, I think, evident that Dr. Stirling's otherwise very clever and interesting tract, "As Regards Protoplasm," is really not directed against the true protoplasm theory as long ago expounded by Fletcher, but only against certain omissions and ambiguities in Mr. Huxley's mode of presenting it to the public.

It may be interesting to trace back a little farther the source of these ambiguities, as I think it will thereby be made plain that unless we frankly admit Fletcher's doctrine of the non-existence of the so-called proximate principles—albumen, gelatin, fibrin, &c.—in the living matter, we cannot maintain the theory of vitality as a property of protoplasm. Now Mr. Huxley makes no pretension to originality in expounding the protoplasm theory—in fact, he distinctly disavows that. But instead of going back to Fletcher, who first put it forward as an hypothesis, or to Beale, who discovered it in its complete form, he takes Häckel as his guide. Let us see how far Häckel is qualified to be a safe and unprejudiced guide. Here, however unwillingly, we are compelled to touch on doctrines of general philosophy which trench upon revealed religion. The fact is that Häckel is a Pantheist, or what we are accustomed to look upon as

an Atheist of the positive kind, although he objects
to the appellation atheist, and "hurls back the severe
reproach" upon the stupid and ignorant people who
apply it to him, declaring that his Monism, or Pantheism, is the only pure Monotheism.*　Not to dispute
about words, it is enough to say that he holds that
"all matter is eternal:" creation of something out of
nothing is unthinkable, and the creation by an external
power of organic forms, even the simplest, out of
already existing matter, is untenable (i. 171).　He
denies the existence of a Personal God, or Creator;
with him "God is the universal Causal Law;" "He
can never act arbitrarily or freely—i.e., God is Necessity:" "God is the source of all force and all matter."
He says, "Monism involves the unity of God in the
totality of nature," while the ordinary idea of God involves "Amphitheism," and, in addition, in the belief
of the majority of people "there exist a number of
other Gods, such as the Devil, Angels, and Saints, who
are either worshipped or feared, and thus their amphitheism is stamped as a very Polytheism" (ii. 451).　As
the natural result of these principles, he denies all
possibility of miraculous intervention, all revelation
and all religion founded on a presumed supernatural
revelation.　But in defence of the purity and eleva-
tion of his principles, he says†—"For us all nature is
animated—i.e., penetrated with Divine Spirit, with
law, and with necessity.　We know no matter without
this Divine Spirit, and no spirit without matter." . . .
" While we recognize the unity of the whole of nature

* " Generelle Morphologie," ii. p. 449.
† " Natürliche Schöpfungs Geschichte," 2nd edit., p. xxix.

and the Divine Spirit everywhere acting therein, we
lose, it is true, the hypothesis of a Personal Creator,
but we gain instead of it the undoubtedly far more
elevated and perfect idea of a divine spirit penetrating
and filling the universe. According to our conviction,
this idea (consistently carried out!) is alone able to
reconcile the still existing opposition between Realism
and Idealism, Materialism and Spiritualism, and to
fuse them together in the far higher conception of
Monism."* And if people will call him a pantheist

---

* There is a vague grandeur about this sentence which is imposing
at first sight, but, on closer inspection, there is little real meaning in
it. For what does he mean by spirit and spiritual? In explanation
he quotes from Goethe, " Matter can never exist and be active without
spirit, nor spirit without matter." Now this appeal is quite useless, as
the true nature of force was not known to Goethe, and the meaning
of his "spirit" is probably a vague compound of what is really force,
and of an allusion to the immateriality of every action, including life
and mind. Häckel also appeals to the authority of August Schleicher
(" Die Darwinische Theorie und die Sprachwissenschaft," 1863), from
whom he quotes a sentence containing these words—"There exists no
matter without spirit (without the necessity determining it), but just
as little does there exist spirit without matter—or rather, there exists
neither spirit nor matter in the ordinary sense, but only one, which is
both at the same time. To brand this view, which rests on observation, with the epithet materialism, is just as perverse as it would be to
impeach it as spiritualism." Here, we perceive, by that little paren-
thesis, that "spirit" means simply *the inherent properties of matter.*
Truly, there is nothing new under the sun; but who would have
thought of a revival of the " essences" of the ancients as a way for the
modern German Pantheists to get out of the reproach of materialism?
Häckel's own use of the word spirit (Geist) in other places is equally
fatal to the apparent meaning of the above paragraph. For instance,
the expression, " We know no matter without this *Divine Spirit,* and no
spirit without matter," he explains away in the " Generelle Morphologie" (bk. ii. 449) thus : " We know a spiritless matter—*i.e.,*
matter without force—just as little as an immaterial spirit—*i.e.,* a force
without matter." His idea of "spirit" is, therefore, simply " force,"
and he comments on the incongruity and absurdity into which people
fall when they attempt to imagine an " immaterial force," such as
spirit, soul, vital principle, creative force, and so forth, for, in fact,
what they picture to themselves is, after all, something material, such
as gas, imponderable matter like heat, or light, or the æther, &c. And

and heretic, he consoles himself with the thought that he suffers in company with G. Bruno, Spinoza, Lessing, and Goethe.

Far be it from me to insinuate anything to the prejudice of those who have from one cause or another shaken their minds loose from the dogmas of the Christian religion, and to question their conscientiousness and love of truth. On the contrary, they are to be honoured in having the courage to express what are, no doubt, their sincere convictions. But with these doctrines, which he presses forward in an enthusiastic and almost fanatical spirit, we cannot but see that Häckel must prejudge the question of the origin of living matter. He accepts the self-existence of matter and force from eternity, and our first knowledge of them is in the form of the " gasiform chaos" of Laplace. Then, as living matter could not exist in our globe till after it was sufficiently cooled down

---

he ridicules the idea of a Personal Creator as an immaterial being or spirit, which last is commonly thought of as a kind of gas or æther (i. p. 173). Nor will the above paragraph in the text look any better if we translate the word " Geist" as mind, instead of spirit. The word " divine" would therefore apply to a kind of infinite mind, bearing a similar relation to the interactions of the matter and force of the universe as our minds do to our brains. But this mind, according to Häckel's own showing, could not do anything, or have the slightest influence on the phenomena or events of the universe, nor would it apparently be conscious—truly not a sublime idea of God! But even this is not consistent with his own showing elsewhere, for he denies the existence of thought anywhere except as the attribute of the matter of the brains of the higher vertebrates, and combining with this the common idea of mind and spirit in the being of God, he says, " We thus reach the paradoxical conception of a gasiform vertebrate— a *contradictio in adjecto*" (i. 174). I hardly think we can be accused of injustice if we regard this Monismus or Pantheism as indistinguishable from Atheism.

to contain liquid water, it must have had a beginning. It is, therefore, of the last importance for his doctrines that primitive generation of living from inorganic matter (Archigony) should be proved possible by natural causes. This is a necessity shared by all schools of philosophy which deny the possibility of miraculous interference at any stage of the world's history, and a strong bias is given to reduce to a minimum, and if possible obliterate, the radical distinction between dead and living matter, and to show how one may pass by insensible degrees into the other. We all know how the presence of a strong wish or prepossession of any kind warps the mind, and insensibly perverts the reasoning powers. This is conspicuously shown in the present instance, where, by constantly speaking of the living matter as albuminous and the like, Häckel has come to persuade himself that the gulf between the organic and inorganic kingdoms is not so great after all. At any rate, he has succeeded in throwing a veil of ambiguity over the subject, which has produced the misconception of the protoplasmic theory above noticed. In proof of this, I may give a few examples. Speaking of the ovum, he calls it " a little lump of albumen, in which another albuminous body is enclosed—the nucleus."* " The nucleus, we can imagine, may arise from purely physical causes by condensation of the innermost central particles of albumen " (p. 306). The Monera are "simple individualized lumps of albumen."† These expressions are constantly used, and he speaks also of

* " Nat. Schöpfungs Geschichte," 367.
† " Gen. Morphologie," i. 182.

viable [lebensfähig] plasma, that is to say, any quaternary compound like albumen which is ready to individualize itself and become living. This corresponds to the organizable matter of many physiologists, and all believers in spontaneous generation. It is needless to say that the term organizable is utterly unwarranted by any facts. The very use of such an expression involves an hypothesis, which is immediately converted into an assumption and argued upon, while it is forgotten that there is not a shadow of proof that albumen is any nearer living matter than carbonic acid or ammonia, and that all we really know is that it is a pabulum more easily assimilated than the binary compounds by certain organisms, although it is decomposed in the process. This is certainly not being organized in the sense used above. However, for Häckel it is not difficult to imagine [Autogony] that in the primeval world, under such different conditions of heat and moisture, &c., ternary and quaternary compounds were formed constituting the "Urschleim" (primordial slime or mucilage). In this "viable plasma," *Plasmagony* takes place thus: "The first organic atom group, perhaps an albumen molecule," attracts other similar atoms in the mother-liquor, like the nucleus crystal. Thus the "little granule of albumen" grows and forms itself into a structureless Moner ("Gen. Morph.," i. 181). Thus the constant tendency is to identify the living matter with the proximate principles we find after death, and to keep in the background and glide gently over the vast and irreconcilable difference between growth and development, and all possible functions of any kind of merely

17

chemical combination.   No doubt, to objections that
the living matter must surely be different from albumen, Häckel will reply, with Huxley—of course it is!
and will quote passages where he calls it albumenoid,
or albumen-like, or, as in the passage here given at p.
247, attributes to the plastids an infinite variety of
atomic constitution.   But the passages I have quoted
above are nevertheless ambiguous and certain to mislead others, and, I maintain, show a confusion in the
mind of the author himself, which must affect all
materialists who do not follow Fletcher's theory of the
metabolic state (see p. 182).   Finally, however, Häckel
is compelled to admit that "Autogony (and also plasmagony—the other form of Archigony) remains a *pure
hypothesis*, because we take for granted a natural process, the transition of lifeless matter into living matter,
which has never yet received an empirical foundation
by trustworthy observation " (" Gen. Morph.," ii. 292).

I do not know what Mr. Herbert Spencer's religious
opinions are, but he, like Häckel, is a follower of Oken
and Lamark in believing in the evolution of living
from inorganic matter by natural processes.   That is
to say, while he brushes away with a kind of scorn
what is usually called spontaneous generation, viz.,
the evolution of " creatures having quite specific
structures in a few hours without antecedents calcu-
lated to determine their specific forms," in fermenting
and putrefying fluids : yet he holds that the process
did take place in the primeval world, and might
possibly even now take place just as Häckel describes
above, by gradual steps.   There never was a " first
organism," but a gradual formation of more and more

complicated chemical products, till finally " organizable
protoplasm" was reached. This expression, of course,
I regard as even more objectionable than Häckel's
viable plasma, and, in fact, is either absurdly tauto-
logical, or meaningless. He gives as a specimen of the
mode in which " organizable protoplasm" may have
been reached in the laboratory of nature, an analysis
of the process by which a complex substance, the
butyrate of dimethylamin, can be made in the laboratory of the chemist. Mr. Spencer describes this process in a somewhat general way, giving the series of
products obtained by successive substitutions and reactions; but as this hardly exhibits the nature of the
process in its true light, Mr. Edward Davies, Chemist
of the Royal Institution of Liverpool, has gone over
the matter in full detail. I give the description in
detail, because although the extreme difference between living matter and the proximate principles
which are the products of its death, has been insisted
upon over and over again ; yet nothing can give an
adequate idea of that except reiteration of details :—

" The following is a sketch of the simplest process known to
chemists for obtaining butyrate of dimethylamin by the use of
inorganic materials alone :—
" Firstly, we pass bisulphide of carbon and sulphuretted
hydrogen over copper at a red heat, and get light carburetted
hydrogen. This is then treated with chlorine, to give chloride
of methyl, and this, with caustic potash, furnishes methyl-
alcohol. Iodide of methyl is made by distilling phosphorus,
iodine, and methyl-alcohol together. The iodide of methyl is
heated with ammonia in a sealed tube, when iodide of dimethylamin is formed with other products. The iodide of dimethylamin, distilled with caustic potash, gives dimethylamin.

" The butyric acid is made by combining carbon and hydrogen directly, by forming the electric light between carbon points in an atmosphere of hydrogen, thus making acetylene. The acetylene is absorbed by ammoniacal solution of cuprous chloride to form cuproso-vinyl oxide. This compound, heated with zinc and dilute ammonia, yields ethene, or heavy carburetted hydrogen. This, absorbed by strong sulphuric acid, gives vinyl-sulphuric acid, which, diluted with water and distilled, gives ethyl-alcohol. Alcohol is converted into iodide of ethyl (in the same way as iodide of methyl is formed), which is heated with zinc in a sealed tube, when quartane, or $C_4 H_{10}$ is formed. Chlorine converts this into quartyl-chloride, $C_4 H_9 Cl$. This is heated with potassium-acetate and strong acetic acid, which forms quartyl-acetate, and this, treated with barium hydrate, yields butyl-alcohol. The acetic acid used in this reaction can be made by oxidizing ethyl-alcohol by platinum black. The butyl-alcohol, by oxidation with platinum black, gives butyric acid. The butyric acid has now to be mixed with the dimethylamin, to give the butyrate required" (MS. Letter).

Thus we have an enormously complicated process, requiring numerous changes of temperature and a carefully-watched and designed succession of processes in an order which, if interrupted or left to chance, would spoil the whole. Does any one really feel that his comprehension of the formation of complicated products by chance at the bottom of the sea, or in the neighbourhood of volcanoes in the primeval world, is really helped by looking into these details? Surely the contrary! And when you have got this substance, you are as far on your way to albumen as a man ascending a small hill would be on his way to the moon. And when you have got albumen, you are still as far from living matter as in the moon you would be from the fixed stars, for all we have yet had proved. Mr. H. Spencer trusts chiefly, like the rest, to minimizing the distinction between the living matter and albumen and protein, thus hoping " to bridge over the interval" between them. The consequence of this is, that he speaks of albumen as taking part in truly vital processes, and in reality does not draw a line distinctly between those physical and chemical processes occurring in living bodies and the truly vital ones. He speaks of the commencement of living matter with portions of protoplasm "less distinguishable from a mere fragment of albumen than even the protogenes of Häckel" (" Biology," 481); and again, that the said protogenes is " distinguishable from a fragment of albumen only by its finely granular character" (" Psychology," 137). With such expressions, how can the public be otherwise than bewildered, and remain unable to understand and appreciate the protoplasmic theory ? We can understand now how Mr. Huxley, when expounding the doctrine of these teachers, should fail to convey a clear idea, more especially in oral teaching, and probably pressed for time. I have no reason to suppose further, that he sympathizes with Häckel's religious opinions ; but he would be naturally unwilling to enter on such a subject in that place.

The above ambiguities and obscurities on the part of Mr. H. Spencer are evidently due to his overmastering desire to bring in the hypothesis of evolution from inorganic matter. Nevertheless, he was obliged to fall back upon the theory of physiological units, as we have seen (p. 182).

But we have a right to protest against the too common practice of naming Oken, Lamark, Herbert

Spencer, Häckel, and others, in the same category with Darwin, as "evolutionists." The four first-named hold the purely speculative opinion of the origin of living from inorganic matter by natural processes. No such processes can be proved to exist now, nor is there any evidence that they ever did exist, which does not beg the question at issue. Evolution, in such a sense, is therefore a pure speculation, derived from other sources than biology, and has no title to be placed on a level with the legitimate theory of Darwin, who, like a true philosopher, founds his explanation of the origin of species on natural processes, viz., variation, heredity, and natural selection, which do exist as a matter of fact. Where these end, he stops and refers the origin of life itself to the miraculous interference of the Creator, who "breathed the breath of life" into certain primitive stem-forms of beings.

It is hardly necessary to say that the so-called spontaneous generation of the putrefactive infusoria, which has lately attracted so much attention, may be put in the same category with primeval plasmagony, as indisposing its adherents to admit the strong distinction between albumen and protoplasm. To discuss this would require a volume, so I may merely state that I have followed closely all the published evidence on both sides, and have contributed some original observations. As far as my opinion may have any value, I may state that, as the question stands at present, the whole evidence in its favour rests on the appearance of bacteria in solutions which have been exposed to a heat sufficient to kill all visible adult bacteria, while it is assumed that any presumed invisible germs of that creature must also be killed by the same agency. To this it is replied that, 1st, Owing to various circumstances, especially the presence of solid particles in the solutions, the heat really did not penetrate to all the individuals in the time during which it was applied; and 2nd, It was shown by the observations and experiments of Mr. Dallinger and myself * that, in the case of several monads found in putrefying solution having the same title to be called spontaneous products as bacteria, besides their mode of multiplication by fission, they were propagated by extremely minute germs, which were seen to grow to the adult form after exposure of the slide to a heat which was fatal to any adult living creature. . I hold, therefore, that, although the question is not yet fully settled, the adherents of spontaneous generation of putrefactive infusoria have entirely failed to establish their position.

Nevertheless, I admit that the recent investigations of the subject, especially by Dr. Bastian, in his " Beginnings of Life," are worthy of all praise, and it is consistent with sound philosophy that the question should be submitted to experiment. It was not a matter to be cried down, as some have wished to do, but it is a subject in which repeated experiment was necessary, and is still necessary.

Nor do I agree with those who condemn the whole object of the investigation as impious. Those who suggest that it is so should remember that Needham, who revived the doctrine in the last century, was a Catholic priest, and was, nevertheless, free from any

* " M. Microscop. J.," 1873-4.]

imputation of heresy. The truth is, it is no more necessarily an atheistic doctrine than the reference of physical phenomena to their natural causes, nor in biology than the theory of the origin of species. It does not necessarily deny creation or design, but it merely pushes the visible interference of the Creator's hand a stage farther back. The whole course of science is an illustration of the substitution of natural processes for the immediate interference of the Almighty as the cause of the phenomena of nature. In the present instance no natural process has been discovered which can explain the origin of living matter, and we may depend upon it that if such be discovered, it will not tend to diminish in the least the enormous gulf that exists between the actual living matter and that which is not living.

CHAPTER XII.

WHEN we take to pieces the materials of a house, and scatter them, the individuality of that house is lost for ever, or remains only as a memory or a history. In like manner if we decompose, or even evaporate a drop of water, its individuality is lost. But with both these, if the materials were preserved, the house or drop might be set up again as before. But with the organic creation, if the life of the individual consists in the arrangement of its particles alone, destroy that, and you destroy it for ever, for had you even the component particles, to the last atom in the exact proportion, you could never set them together again and reproduce the individual. On the protoplasmic theory of life death is death for ever, as far as science alone can teach us. And for some, who will accept no teaching but the teachings of science, death is death for ever, of body, mind, and soul, for man as well as for all the rest of the animated creation. And thus the materialist theory of life and mind comes to have a meaning which excites emotions of the deepest horror and aversion. We must not allow our natural

feelings to overpower our judgment, so as to hinder us
from looking at this question steadily in the face;
when we may perceive that the word materialism does
not necessarily bear all the meaning that is usually
put upon it.   There is materialism and materialism.
With Häckel it bears the extreme signification of the
denial of the immortal soul of man as well as of a
spiritual principle of mere life, and when he is re-
proached with materialism he disclaims it in the
ethical and æsthetic sense, viz., as a low tone of mind
which is set upon money and material possessions, the
gratification of personal wants ministering to bodily
comfort, to luxury, vanity, and other objects of vulgar
ambition, in preference to the more spiritual states
desiring intellectual, artistic, and the higher moral objects involving self-sacrifice.   And he points to the
fact that as a rule philosophers of his stamp are men
of higher intellectual and moral aim, and of more
virtuous lives, than many professedly religious persons.*
This is, no doubt, the fact, and probably it is all that

* " Hence," says he, speaking of the low ethical materialism, " you
will seek it in vain among such materialists and philosophers whose
highest delight is the mental enjoyment of nature, and whose highest
object is the recognition of her laws. For this materialism you must
look in the palaces of the princes of the Church, and in the conduct
of all those hypocrites who, under the outward mask of pious honour-
ing of God, strive solely for hierarchical tyranny and the making material gain out of their fellow-men. Dead to the infinite nobility of
the so-called ' brute matter,' and the splendid phenomenal world
springing out of it ; insensible to the inexhaustible charms of nature
as they are ignorant of her laws, they brand as heresy the whole of
natural science, and as sinful materialism the culture springing from it,
while they themselves revel in the most repulsive form of the latter. Not
only the whole history of the popes, with their endless chain of terrible
crimes, but also the repulsive ethical code of the orthodox in all
forms of religion furnish sufficient proofs on this point" (" Natürliche
Schöpfunggesch.," p. 33).

is to be said on his side of the subject, and it is at any
rate plain and straightforward.   But the treatment of
the subject by Mr. Huxley is perplexing in the extreme.   After setting forth the inevitable materialism
of the protoplasmic theory in the plainest manner, he
suddenly halts and declares that he is no materialist,
but, on the contrary, he believes that materialism involves a grave philosophical error.   In explanation, he
gives several pages of metaphysical reasoning on
Hume's doctrine of causation and Berkeley's idealism,
which, it does not require Dr. Stirling's trenchant
criticism—common sense is quite sufficient—to show
us, leaves this matter exactly where it was.   The question is simply evaded.   Perhaps Professor Huxley's
motive for this may be the natural aversion felt by
men of taste for bringing their religious opinions before
a mixed audience, and also that he is mindful of the
counsel of Bacon :—" If we were disposed to survey
the realm of sacred or inspired theology, we must quit
this small vessel of human reason and put ourselves
on board the ship of the Church, which alone possesses
the divine needle for justly shaping the course.   Nor
will the stars of philosophy, that have hitherto principally lent their light, be of further service to us ; and
therefore it were not improper to be silent upon the
subject."
The same repugnance was felt by Fletcher, but he
felt also that we are reminded by a far older and wiser
authority than Bacon, that if there is a time to be
silent there is also a time to speak, and that it became
his duty to speak plainly.   Accordingly he does not
hesitate to say it is materialism, absolute, blank mate-

rialism in the ordinary common-sense meaning of the term without any metaphysical subterfuge in the background, as applied to the scientific doctrine of life and mind of animals and of man, in as far as these attributes are common to him and them. But what of this? Is it not the veriest truism, and the mere expression of the fact of mortality common to all living beings, including man? "For when the breath of man goeth forth, he shall turn again to his earth, and all his thoughts perish" (Ps. cxlvi.). What further difficulty is thrown into the conception of the resurrection of the body when we have always known that the elements are utterly dispersed and used over and over again in numberless different organized individuals? The miracle of resurrection stands exactly as it was, whether we attribute mortal animal life to a spirit—to us incomprehensible—or to a collocation of material atoms. It is true that the materialism of the protoplasmic theory has been used as an argument by those who on other grounds are disposed to reject all supernatural interference and revelation; but then such persons will seize on every fact and doctrine in natural science which may be made to appear to serve their purpose. No better example of this can be found than that of Voltaire, who, ignorant of the development that geology was about to undergo, attempted to explain away the significance of the shells on Alpine mountains by their having fallen from the hats of pilgrims, when their presence was adduced by equally ignorant religious partisans as evidence of the deluge. In allusion to these perversions of the foregoing theories of life and mind, Fletcher says—

"It is lamentable that the last, and certainly most philosophical view of the question, should have been allowed by the intemperance of some of its partisans to involve influences which have rendered the word materialism hateful to all pious men, and gained for its advocates the title of philosophers, at the expense of one which should have been much dearer to them" (iii. p. 95).

And the grand principle by which the legitimate and scientific meaning of materialism is separated from the unwarrantable use of the word as synonymous with religious infidelity, is simply the strict separation of the ideas of life and mind from that of the immortal soul, and the humble acknowledgment of our sole dependence on miraculous revelation for all our knowledge of what appertains to the future life of man. I cannot do better than quote the whole passage from Fletcher in explanation of these views :—

"Nor is this view of the matter, as is sometimes vaguely supposed, in any degree hostile to, or inconsistent with, the purest and loftiest religion. The hackneyed arguments against this opinion, founded upon its supposed immoral tendency and impiety, appear to proceed upon the principle, certainly erroneous, that the mind and soul are identical. Who, that has watched for five minutes the action of a dog, can be so blinded as to deny that he possesses attention, imagination, abstraction, judgment, desire, grief—in short, all the intellectual faculties and passions, in the display of which *thought* consists, but who will attribute to him an immortal *soul?* The existence of such a substance, attached during life to the body of responsible man, and surviving him to all eternity, we are at once intuitively led and explicitly taught to believe ; but it is a question of morality and faith, not of physics and demonstration, and to be determined not by its susceptibility of proof, but at once by its verisimilitude, and by our confidence in the authority on which it rests.   Who that has contrasted, as every one does, and must do, the chaotical condition of the moral world in this state of our existence, with the harmonized operation of the physical, if we believe that they are equally directed by the same Almighty hand, can avoid believing (and the belief is, therefore, and has always been, almost universal) that this Almighty hand has set apart His own time for rectifying this inconsistency ; and when, in addition to this intuitive persuasion, we have the assurances of revelation to the same effect, what need have we to look to physical philosophy to shake our confidence in its truth ?

"The two subjects seem to be utterly unconnected. 'I have no hope of a future existence,' observes the late talented Regius Professor in the University of Cambridge, 'except that which is grounded on the truth of Christianity;' and it was well remarked lately, 'that if man be not satisfied to place his hopes of immortality in a Divine gift, he must confess that the difference between his own claims and those of many other animals is in degree only, and that degree in some instances a very small one.'   That the soul is something absolutely distinct from mind, which is *nothing*—or, at least, *nothing substantial*—cannot be doubted; but what the nature of the soul is, it will

be time enough to begin to investigate when we can
conceive the nature, distinct from the properties, of the
least of the particles entering into the composition of
one of the filaments of the down upon a blade of
grass; though even when we have succeeded in perfectly comprehending this, and much more than this,
what right have we to presume that nothing *can*
exist which is beyond the sphere of our comprehension ? The nature of the soul is probably such as
man in his present state has neither words to describe
nor faculties to understand. His efforts to do so, like
the attempts of one born blind to conceive and describe
the nature of light, are perhaps as unreasonable in
their object, as they have hitherto been unsuccessful
in their result; and, for aught we know, a true sixth
sense (for who shall say that every possible form of
sense has been in man exhausted ?), with all the new
ideas which would thus rise, may still be necessary
before it can be comprehended and expressed. What
would be the consequence of a further insight—
whether it would assist us in our duties, or divert us
from the performance of them, is very uncertain.
The withering and impious inference, therefore, which
has sometimes been drawn of the mortality of the soul
from that of mind, is as totally unwarrantable on the
one hand as the whining and canting exception which
has been so commonly taken to the mortality of the
mind, from the supposed necessity of that inference.
We cannot conceive, it is said, the nature of the soul
distinct from the mind. God alone knows how little
the most profound of us, big with the conceit of penetrating into the sublimest mysteries of His greatest

works, really and truly knows of the most familiar
features of the least of them; and God, it is to be
hoped, will pardon as well as pity (for He made man
daring as well as imbecile) at once the rash flippancy
with which the firmest and best persuasions of natural
reason and the most sacred doctrines of Revelation
have been braved, because they have *appeared* to be
incompatible with philosophy, and the bigoted blind-
ness with which the most evident deductions of philosophy have been spurned, because they appeared to
be opposed to natural reason and revelation.   As often,
then, as it shall be said that mind, or the faculty of
thinking, is a property of living matter, as much as
irritability or sensibility are properties of it—that it
is born with the body, is developed with the body,
decays with the body, and dies with the body—it is
understood to be the mind only, not the soul.   The
soul is certainly something not material indeed, but
substantial—a divine gift to the highest alone of God's
creatures, responsible for all the actions of the mind,
but as totally distinct from it as one thing can be from
another—or rather, as something is from nothing"
(iii. 93, 94).

Further comment is superfluous on this point, but I
may add that Fletcher's opinions were in favour of the
physical rather than the teleological view of the nature
of things.   He did not reject altogether the appeal to
final causes, but held, that although all things were
created with fore-knowledge and purpose by the
Almighty, yet His will was carried out by the interaction of ordinary matter and force in blind obedience
to the properties originally impressed upon them, both

in the inorganic and organic worlds, and not by any intermediate agencies specially appointed for the particular end.

By the absolute distinction of the soul from the mere life and mind of man, those trained in the school of Fletcher have been able to follow with equanimity the complete change that has taken place of late, especially in this country, in the abandonment of teleological views even in biology. The difficulty of accounting for the origin of species and the millionfold examples of exquisite adaptation of structure and function to the apparently designed purpose displayed in the organic kingdoms, had long been the stronghold of the argument from design, but since the theory of Darwin the whole doctrine has received so rude a shock that it may be said the time is come when so-called natural theology must be banished from the sciences. Darwin's theory is a theory properly so called, and not an hypothesis. It does not rest on any new postulate as to the powers of organized matter not already known to us by sensation and experience. On the contrary, it is founded on facts already established, viz., heredity and adaptation. It is, therefore, not wonderful that it should have been so rapidly accepted by all philosophers as the efficient cause of the origin of species and the building up of persons, or individualities of the higher orders, from independent vital units or plastids of lower orders of individuality.*   No doubt there are

* That the unity and personality of the animal body is due to a spirit, or *anima*, proper to each individual will hardly be maintained by any one now. It will not be said that an *anima* proper to each constitutes its personality, and that this *anima* secretes *its* bile through its liver, or circulates its blood by means of its heart. On the contrary, it

18

a great number of instances in which it cannot yet be shown to be the sufficient cause. But these are merely difficulties in the application of the theory, and although the anti-Darwinians rest their reputation as philosophers on these difficulties, yet it is impossible for us to rest our belief in the being and attributes of God on such a foundation as the residuum of unexplained phenomena in the world of biological science at any particular time, and which is diminishing daily by the progress of knowledge. We must therefore abandon at once the argument of design furnished by the organic world as anything superior to, or differing in principle from, that furnished by the inorganic, and the teleological significance of the former must be reduced to no greater, but also to no less, than that of the latter. For long the expansion of water on freezing was brought forward as an instance of beneficent design, being an apparent exception to the general law of contraction of all bodies by cold, for the purpose of preventing lakes and rivers from freezing to the bottom, whereby a large part of the globe would have been rendered uninhabitable for man. But since it has been found that bismuth, iron, and several other

---

will be readily admitted that each individual is composed of a number of vital units specifically different, built up in harmony by certain laws of germinal development, and maintained in harmonious action by the ties of a common circulation and nervous system. It is otherwise, however, with the Mind, and, as yet, few are prepared to admit that the unity and consciousness of personal identity of each animal can be produced without a single essence or immaterial principle, which, it is said, works through or by the separate cerebral faculties. Nevertheless, however little we can, as yet, understand or explain how this unity is produced by the harmonious working of an extremely complicated congeries of separate parts, yet we must admit that it is the fact; for this sense of unity and personal identity may be impaired, or even lost, like other mental faculties, by disease and decrepitude.

substances expand on solidifying, the argument from ice has lost its significance as a special exception.

In like manner, in the organic world, the very existence of parasites exquisitely contrived to prey upon, and often destroy, even the highest of organized beings, deprives the argument for design of much of its significance, but the following circumstance reduces it almost *ad absurdum*, namely: it is found that two species of tape-worm could not run through their cycles of generation, except by abiding for one phase of them in the human body, which is the only one in the whole creation fitted for them, and without which the said tape-worms would cease to exist.   Are we to infer, says Dr. Cobbold, that man, who is so exceptionally adapted for the purpose, was contrived for the sake of the worm ?   In like manner, the same difficulties and inconsistencies may be found to run through the whole argument from final causes. It would appear that, in spite of the well-meant endeavours of many estimable persons, it savours more of presumption than reverence on the part of finite human creatures with such a limited sphere of knowledge, to attempt to climb to the throne of Omnipotence by the ladder of inductive reasoning, and to discover the design and purpose of the Almighty ; or to find out His nature and attributes by the only kind of reasoning which human beings are capable of, viz., that founded on sensation and experience.   On the contrary, the more knowledge extends, the farther we are obliged to put back the inference of the direct interference of a personal Creator, and the less are we able in analogy with our own nature to argue the being and attributes of God from the evidence of specific design and purpose. In this sense, also, we may accept the declaration of God as a jealous God, for He has shrouded the direct working of His hand in darkness impenetrable to the eye of science.

It is true there still remain two points on which the evidence of science contradicts the assumption that no change has ensued in the world's history beyond what can be traced to existing causes. The law of dissipation of energy—leading to the inference that as the present state of our solar system must have an end, so it must also have had a beginning— and the origin of life on our planet are held to demonstrate the interference at some time of an external power other than those of existing matter and force, of which we are cognizant by sensation and experience. But it is not impossible that even these may be brought within the province of natural causes, so I for one am quite willing to go at once to the extreme, and abandon all pretension to discover from science alone, not only the attributes, but the very existence of God at all, and to rest my whole belief on revelation. Without revelation natural theology has hitherto with peoples always led to Polytheism, and with philosophers and men of science to Atheism or Pantheism. Once the knowledge of a single personal Creator has been received through revelation (in fact Christian writers on natural theology all start from a foregone conclusion to that effect, whether consciously or not), our position towards the invariable laws of nature and to final causes becomes clearer. And we feel that common sense shows no difficulty in the way of the belief in miracles; surely the power which made all things may again, at any time, create or annihilate force or matter and interfere with natural laws at His pleasure!* With respect to final causes we may still, surely, in harmony with our own ideas, admire the glory, the beauty, the beneficence, and the exquisite fitness of things displayed in the universe, although we cannot trace the direct working of the hand of the Creator in the details of the organic world, any more than in the revolution of the planets, nor the shaping of each coast-line or mountain-chain.† Not that I admit that the

* "Operating by intervening laws has exclusive reference to *us*, and is not absolutely necessary. It is evident if the proposed physical theory be true, that the laws are constant only so long as the qualities of the æther and of the atoms on which they wholly depend are constant; and as we have concluded that these qualities were made such as they are by an immediate exercise of power, by power similarly exercised, they might be changed in any manner, and even annihilated. In fact there have been well attested occurrences in the world, which can be accounted for only as being caused by power thus operating. They have been called *miracles*, wonders, apparently on account of their infrequency; but, essentially, they are only repetitions of creative acts of the same kind as those whereby the elements of the world were originally called into existence. Of course, according to these principles, the possibility of miracles cannot be disproved by physical science" (Challis, "The Mathematical Principles of Physics," p. 106).

† "By being conscious that strength and skill are required for making anything, we can understand that these qualities were necessary for the creation of the world, and, consequently, that it might have been created, among other purposes, for that of demonstrating the power and wisdom of the Creator" (p. 105).

"It is only by slow degrees and other prolonged intellectual labour, that human intelligence has in some measure succeeded in deriving the laws from the original conditions; whereas, by the Supreme Intelligence, all such consequences must have been intuitively seen when the conditions were first imposed. Thus the theoretical study of physics is specially adapted to exalt and give distinctness to our conceptions, both of the wisdom and the power of the Creator. On the principle of final causes it may be asserted that, together with many other purposes, this was contemplated in giving to the æther and the atoms their specific qualities. If I thought otherwise I should not be able to consider as justifiable the devotion of many years of my life to physical researches" (Challis, p. 106).

origin of species by natural laws instead of direct creation, relieves us of the difficulty of the origin and existence of evil and suffering in the animal creation, including man. On the contrary, to the believer in an all-powerful, all-wise, and all-good Creator, these difficulties remain exactly the same as before, not less, but also not greater. And on the whole, we may look upon the microscope, and the telescope, and the progress of knowledge generally, to have simply widened the field of knowledge, but not altered the relations of science to the real doctrines of revealed religion. Whoever is disposed to cavil at these doctrines, need not go to the more recondite truths of science for illustrations. He does not require to search out the cause of the marvellous adaptation of animal species to their habitation and purpose by the cruel and pitiless law of survival of the fittest. He has but to open his eyes and look around and he will see thousands of examples of the same difficulties in daily life. Does not the rain fall equally on the just and the unjust? does not the stone crush and the fire burn the martyr equally with the criminal? does not the wicked flourish while the good may suffer wrong; and so on through every phase of human life? If to the end of time these difficulties will perplex the mind of believers, still there remains the hope that in a future state of existence, beside redress of the balance of wrong and suffering in this world, the purpose of God in causing immortal beings to pass under the hard necessities of fixed natural laws during the first stage of their existence, will be made manifest and reconciled with His benevolence and omnipotence. But to the merely natural theologian, the existence of ugly and noxious creatures, whose purpose is destructive of the beauty and well-being of the rest of the animal kingdom, is fatal to the discovery à *posteriori* of a single infinitely wise, good, and powerful God, after the manner of a scientific problem. And, accordingly, the outcome of natural theologies has hitherto been a plurality of gods, not infinitely powerful, and at war with one another; or else blank Atheism, which escapes inconsistency by abolishing all plan and all creation whatever. Theologians and preachers of revealed religion, who think to excogitate the nature and attributes of God by their own reason and from the evidence furnished by science, may with profit take warning from the grim irony of Häckel,* and go back to what God has revealed of Himself, as the sole source of all our knowledge of His nature.

For myself, I am content to believe in no God, angel, or spirit, or the immortal soul of man, except as made known to us through the miraculous specific revelation contained in our Scriptures. At the same time, these beings are of a nature to us wholly incomprehensible and inconceivable. The cardinal doctrines of revealed religion are thus dogmas, not resting on any proofs derived from observation or science at all.†
These dogmas are also mysteries, not only incapable of scientific proof or disproof, but also above and beyond the comprehension of the human intellect. Are we,

---

* "The priests say, 'God created man in his image.' It should rather be 'Man creates God in his image,' or as the poet expresses it, 'man paints himself in his Gods'" ("Gen. Morph.," i. p. 174).

† In the language of the man of genius and statesman, Mr. Disraeli, "Where knowledge ends religion begins."

then, to accept blindfold all that has been taught in
each age in the name of orthodox religion? By no
means, for I am content to accept the rule laid down
by many great and good men, that, although we are
bound to receive many things above and beyond our
reason, we are never called upon to believe anything
contrary to reason and the evidence of the senses.
And as it is impossible that what is truly the word of
God can conflict with the works of God or the teachings of natural science, so, in the interpretation of the
Scriptures, science must be the final arbiter as to the
meaning, for all true science is founded on sensation
and experience. Who would be satisfied with the
judgment of a person wholly ignorant of history who
presumed to interpret prophecy? And how does this
differ from permitting persons ignorant of science to
interpret Scripture where it appears out of harmony
with the facts of nature? The accepted interpretation
of doubtful passages of Scripture was fixed by men
like ourselves, perhaps not better morally, and certainly
immensely less qualified by knowledge of the laws and
phenomena of nature.

One cardinal point in dealing with the question of
materialism in this chapter has been the complete abandonment of the argument from design as even a colla-
teral support of revelation. My opinions on this subject were chiefly derived from the conversation and the
written works of that man of genius, Samuel Brown,
whom, in his too short life, I had the happiness to count
among my friends. I think, therefore, that I cannot do
better than conclude this work by giving, in his own
words, a statement of the argument which was published

in the palmy days of Bridgewater-treatise-dom, before the Darwinian theory was promulgated. Dr. S. Brown takes as an illustration of the argument from design the law of diffusion of gases and the manifold useful purposes which it serves in the economy of nature, and proceeds as follows :*

" A bubble rises from the bottom of a solitary pool, basking in the sun among the hills: clothed about with a slender fibre of rainbow-hue, the bonny bell floats like a thing of light over the mantling ripple of its little sea, till the tiny craft is broken on the flower-bud of a water-lily; and away fly its crew of dancing atoms hither, thither, and everywhither !

> " ' A timid breath at first, a transient touch ;
> How soon it swells from little into much !'

" What a wondrous combination of means and ends ; how remote the instrumentality employed from the effects produced; and how worthy of a God ! Suppose, then, a million instances like this ; recall to mind the curious cases you have read in Paley and the Bridgewaters: find many more in the records of science, for every page is full of them ; search out the undiscovered multitudes of similar examples in the open book of nature, which is a written strain of the loftiest music from beginning to end: and you have the data which the natural theologians endeavour to generalize. The survey of these crowding facts, like that of every other class of observations, suggests the inexhaustible inquiry of research, How are they to be

* "Lay Sermons on the Theory of Christianity," No. II. The argument of design equal to nothing; or Nieuentytt and Paley *versus* David Hume and St. Paul. By Fidian Analysis. Edinburgh, 1842, Blackwood. This and S. Brown's collected works are now out of print.

understood ?   What is their meaning ?   Where is the
theory ?   The method of inquisition is the same here
as elsewhere : Is there any similar class of facts of
which we *know* the explanation ?   Yes ! there is one.
In the works of human art, the mechanism of a watch,
the construction of a steam-engine, in every product
of art, there are adaptations of one result to another,
completely resembling those which are found in the
world which art attempts to imitate and control.   In
truth, all art consists in the institution of such mutual
relations and such production of effects by fitly-chosen
means.   Now the explanation of this set of facts is
ready beforehand—Art is the product of a designing
mind, and the designer is man.   Accordingly the substance of the Paleyan argument is this : The facts of
adaptation discovered in nature are radically like the
facts of adaptation instituted in art, and the inference
is that they resemble them in origin as well as resulting character—Nature is the product of a designing
mind, and the designer is God.   This is the argument
of design, and it is essentially cumulative in its power,
for the greater the number, and more striking the kind
of evidences of design that can be gathered around it,
the stronger does it appear to become.   I have some
strictures to make upon it, with the sincere hope of
convincing you that, without a previous or simultaneous act of faith, or intuitive belief, conscious, or
unconscious, it is wholly inadequate to the purposes
for which it was constructed.

   "The first is this :—If there be any genuine analogy
between man, the designer of the works of art, and the
inferred designer of the works of nature, it must be
complete, and extensible in the inference to all the

essential characteristics of the known designer, man. How, then, does man design ? By reducing discovered truth to his own uses, and making combinations of natural forms and qualities. He knows the expansive force of steam, as well as the law of latent heat, and makes a steam-engine : he creates nothing. So that the Deity, inferred from evidences of design, does, for all that the argument of analogy makes out, discover truth, apply it to his own uses, and make combinations of forms and qualities. He knows the repulsive force of matter, as well as the law of gravitation, and makes a solar system : He creates nothing.

"This is Hume's analysis, though otherwise expressed, and very differently intentioned ; and its force is irresistible. Hume is the best analyst, as a mere analyst, that Britain has been able to produce. A sincere, and not uncharitable man, he detested the plausible, and never rested till he stripped it bare, and hooted it out of presence. If he had believed in God by faith, I had only need have reiterated his voice ; as it is, you see how searching and indisputable his analysis is, so far as it extends.

" All that the boasted argument à posteriori, as it is called, for the existence of a God, even tends to establish is the existence of a designer, not that that designer is the Supreme, whom science ' falsely so called' is thus ambitious of demonstrating like any other theorem. . . . . . But allow that the inferred designer really spoke the worlds into existence, and He alone, still that creative designer may not be God after all, for Divinity, if proved at all, must be proved to be almighty in power, inexhaustible in wisdom, and boundless in love ; but the universe cannot be proved

to be anywise infinite in the literal sense of infinitude:
it is only indefinitely vast, its magnitude compared
with true immensity being a trifle, for all our tele-
scopes can disclose, and the attributes of its inferred
Creator may be less than infinite in kind and degree.
Whatever is less than infinitude is infinitely less.
This is not God.   Are evidences of goodness, wisdom,
and power of no worth then, and manifestations only
an unmeaning pageantry ?   Is it to no purpose that we
see :

> " ' All things with each other blending ?
> All on each in turn depending :
> Heavenly ministers descending :   .
> And again to heaven uptending ;
> Floating, mingling, interweaving ;
> Rising, sinking, and receiving
> Each from each, while each is giving
> On to each, and each relieving
> Each, the pails of gold, the living
> Current through the air is heaving :
> Breathing blessings, see them bending
> Balanced worlds from change defending,
> While everywhere diffused is harmony unending.'

Has this mazy universe of melody no significance
beyond its own unfathomable beauty ?   Heaven for-
bid !   Once know God otherwise than by discovery,
once believe His Being upon the same foundation as
you believe the existence of the world without your
own personality, and the truth of self-evident propositions, all of which are incapable alike of proof and
refutation.   Once apprehend Him as the Incomprehensible One, ' in whom we live, and move, and have our
being,' and then the world, and all the worlds, are the
sublimest commentary and illustration of His tran-
scending attributes, being, in truth, His uttered word,

still vibrating under the concave of immensity; and the science of final causes becomes the noblest of man's terrestrial pursuits.   This is the method of the book of Job and the Psalms of David, in both of which the Divine Majesty is tacitly understood as being, of course, the Jehovah, or one independent reality, and His attributes are only illuminated by the contemplation of His handiworks. 'Praise ye the Lord from the heavens. Praise ye Him, sun and moon: praise Him, all ye stars of light. Praise him, ye heavens of heavens, and ye waters above the heavens. Let them praise the name of the Lord; for He commanded, and they were created.'* In reality this is the history of every man's process of thought, with whom the argument à *posteriori* has seemed to himself to have been potential.   Not the argument of design, but the argument of design together with unconscious faith in Godhood, has taught men in all ages to behold the Creator in His works.   In fine, the same must be said of the natural theologians themselves.   They have failed to analyze their own process of conviction for one thing; they have been unable to see through their false argument, considered as a mere analytical argument, for another; and then they have always taken the existence of the world without for granted, while they have tried to prove the Being of God forsooth,

* " Canst thou by searching find out God ?" (Job xi. 7).
"Through faith we understand that the worlds were formed by the word of God, so that things which are seen were not made of things which do appear " (Heb. xi. 3).
" For every house is builded by some man; but he that built all things is God " (Heb. iii. 4).
"By the word of the Lord were the heavens made; and all the host of them by the breath of his mouth " (Psalm xxxvi. 6).
" For this they willingly are ignorant of that by the word of God, the heavens were of old " (2 Peter iii. 5).

although the two propositions are alike insusceptible
both of faithless demonstration and sincere denial.
The first writer of note, who stated the argument of
design as a formal proof, was the Dutch mathematician
Nieuentytt, in whose 'Religious Philosopher' is to be
found the original of that classical analogy of a watch,
which was afterwards expounded by Howe, and then
illustrated and enforced with so much perspicacity and
elegance by Dr. Paley.   The last was the first to urge
it with such effect as to secure it a standing in the
world.   His 'Natural Theology' is read by everybody, and is a text-book at the universities.   It has
gone through many editions, even a cheap one for the
people, and is a standard work.   It has lately been
presented anew, under the united auspices of Lord
Brougham and Sir Charles Bell.   Lastly, the late Earl
of Bridgewater has bequeathed the world eight well-
paid treatises, all emulous of demonstrating *Him* 'who
is past finding out.'   I would not drive counter to such
authorities, if I were not convinced that the cause of
Christianity has suffered from these attempts to afford
it external aid.   Their direct tendency is to rob the
religion of faith of its essential character; and this of
'design' encourages those who reject our most holy faith
in the implied conclusion that either God must be to
be found in nature by research, or not exist at all.
Hence come insincere Atheism, idolatrous scientific
theism, and worthless half belief in God.   These are
my motives and defence. . . . Faith and analysis have
to work together on this momentous theme as on every
other; the former to give assurance of divinity, and
the latter to show what Godhead cannot be, and even,
in some little degree, what Jehovah must at least be.

" All idolatry, from the rudest worship of imperson-
ated physical powers among our Saxon forefathers to
the Christianized anthropomorphism (or way of thinking about Deity as a mere somewhat infinite human
being) and the scientific theism of the present day, is
the product of a lifeless faith in Godhood, and an
adequate analysis of His works.   This is what is in-
veighed against by the fidian analyst, St. Paul.   Nay !
so far from being inconsistent with the Scripture, the
sole aim of this sermon is to inculcate the inspired
declaration of the gifted apostle to the nations, that
' *It is through faith we understand that the worlds
were framed by God.*'   David Hume is right, and so i
St. Paul ; and the pauline, or rather the divine, trutl
is incalculably the greater, containing the other together with as much more as the way of God tran-
scends the thought of man.   That ' *The worlds were
framed by God* ' is a surpassing mystery.   A mystery
and therein surpassing.   Known to be an object of
delight by the fullest assurance of belief, and attested
by the broad signature of the sciences, though it needed
no other than the King of Heaven and earth's ; but
infinitely and for ever beyond created comprehension !
The worlds, what are they singing there, in mystic
choir, in the bosom of that holy sky on which words,
' the winged wheels of thought,' are far too feeble to
discourse ?   What have they been made of, and how,
and when, and where ?   Shall they last for aye, floating in the unimaginable æther of immensity, on the
noiseless surge of which they were launched in the
beginning ?   Or are they, in some billion of cycles,
say thousands of years, perhaps a few short centuries,
or another month from to-day, or even one fleetiug

hour from this last look upon their solemn splendour;
are they to pass away at the silent and unutterable
hest of the eternal King of Glory ?   And the King of
Glory, who is this ?—Ah ! they have toyed too lightly
with the Creative Attribute, who have though. to
climb up to its awful sanctuary, where it dwelleth
evermore in the omnific word of Godhood, by piling
stone upon stone in endless erection of a faithless
science of final causation and a great first Cause.
Why Jehovah is not the infinite source, but the source
of the infinite source ; not the first cause, but the cause
of the first cause of all things ; and even that in an
altogether metaphorical mode.   In fine, raised up on a
basis of unconscious scepticism, this kind of natural
theology is a seemly superstructure ; but it hangs on
the air, wavers uncertainly in every wind of doctrine,
and is ready to vanish at the first sound of a bolder
infidelity, leaving no trace behind.   But the same
phantasmagorial temple of design, with its magnificent
proportions, shapely columns, rare devices, and choicest
ornaments, becomes a grand reality, the instant that
man, as its anointed priest, proclaims through the
resounding aisles that FAITH, ONLY FAITH, IS THE
EVIDENCE OF THINGS UNSEEN."

THE END.

LONDON : BAILLIERE, TINDALL, AND COX, PRINTERS.

20, *King William Street, Strand,*

London : January, 1875.

PUBLICATIONS

BY

BAILLIÈRE, TINDALL, & COX.

Essays on Conservative Medicine, and kindred topics.
By AUSTIN FLINT, M.D.,
Professor of the Principles and Practice of Medicine, and of Clinical Medicine, in Bellevue Hospital, Medical College, New York.   Price 6*s.*

Lessons in Hygiène and Surgery, from the Franco-Prussian War.   Forming the most complete text-book of Hygiène and Military Surgery of modern times.
By Surgeon-General C. A. GORDON, M.D., C.B.,
On Special Service, on behalf of Her Majesty's Government, with the Army in Paris.   Illustrated, 10*s.* 6*d.*

" The work is an exceedingly valuable one as a record of personal observation, and of experiences so recently acquired in varied and extended fields of observation, by a medical officer of distinguished position and high professional attainments."—*Army and Navy Gazette.*

" Dr. Gordon's work has already attained a high character in medico-military literature."—*Boston Medical and Surgical Journal.*

By the same Author, 2*s.* 6*d.,*

Life on the Gold Coast.   Being a Full and Accurate Description of the Inhabitants, their Modes and Habits of Life ; interspersed with amusing Anecdotes, Hints to Travellers and others in Western Africa.

I

By the same Author, super cloth, 2s. 6d. ; or, popular edition, paper wrapper, 1s.,

A Manual of Sanitation ; or, First Help in Sickness and when Wounded. Specially adapted as a pocket companion for officers and privates of the regular and volunteer services at home and abroad, in peace and in war.

"Instructions conveyed in clear and intelligible terms."—*The Standard.*

"The official distribution of Dr. Gordon's little manual throughout the forces, would, indeed, be a boon. Volunteers would also do well to purchase and study it."—*The Medical Press and Circular.*

"Contains a great deal of useful matter simply arranged in alphabetical order for ready reference, and might be of great use to others besides soldiers."—*The Builder.*

"Though meant in the first place for army use, it might with advantage find a place in the household, as it is sensible and practical."—*The Graphic.*

By the same Author, 3s. 6d.,

Experiences of an Army Surgeon in India. A Concise Account of the Treatment of the Wounds, Injuries, and Diseases incidental to a Residence in that Country.

Also, price 1s.,

The French and British Soldier. A Lecture on Some Points of Comparison, delivered before H.R.H. Prince Arthur and the Garrison of Dover.

"Clearly and ably written, and may be read by civilians with much peasure and profit."—*The Figaro.*

Short Lectures on Sanitary Subjects.

By RICHARD J. HALTON, L.K.Q.C.P., L.R.C.P. Edin., L.R.C.S.I., &c., Medical Officer of Health to Kells.   5s.

Lecture   I. The Necessity of Sanitary Science.
,,   II. Air.
,,   III. Sanitary Science in the Sick Room.
,,   IV. Ventilation.
,,   V. The Relation of Popular Literature to the Public Health.
,,   VI. Food.
,,   VII. Clothing.
,,   VIII. Cleanliness.
,,   IX. Sanitary Science in Relation to the Training and Education of the Young.
,,   X. Epidemics.
,,   XI. The Influence of Amusements on the Public Health.
,,   XII. Over-crowding.

Food; its Varieties, Chemical Composition, Nutritive Value, Comparative Digestibility, Physical Functions and Uses, Preparation, Culinary Treatment, Preservation, Adulterations, &c.

By HENRY LETHEBY, M.B., M.A., Ph.D., &c.,

Professor of Chemistry at London Hospital, Food Analyst and Medical Officer of Health to the City of London. Re-written, enlarged, and brought up to the present time, 5s.

"An excellent notion of the chemistry and physiological action of the various foods will be derived from a perusal of Dr. Letheby's book, and the reader will be entertained throughout by the narration of many facts which will enliven the study."—*The Lancet.*

"Either as a text-book for schools or as a household guide, it is excellently adapted."—*Public Opinion.*

"Clergymen who are interested in 'cooking for the poor' will find much to aid them in this valuable work, a copy of which ought to be in every household."—*Figaro.*

A Manual of Hygiène, Public and Private, and Compendium of Sanitary Laws, for the information and guidance of Public Health Authorities and Sanitarians generally.

By CHAS. A. CAMERON, M.D., F.R.C.S.,

Professor of Hygiène, Royal College of Surgeons, Medical Officer of Health and Analyst for the city of Dublin. With Illustrations, 10s. 6d.

By the same Author, price 1s.

A Handy-Book of Food and Diet in Health and Disease.

"The newest views as to the physiology of digestion, and the comparative values of food substance. It is short and sensible."—*The Globe.*

By the same Author, 6d.,

On Disease Prevention. A Practical Treatise on Disinfection.

"This little work, which is offered at the trifling charge of sixpence, contains practical directions for disinfecting rooms, clothing, bedding, &c., with chapters on vaccination, water impurities, and other important sanitary matters."—*The Review.*

Also, price 2s. 6d.,

Lectures on the Preservation of Health.

Price 6d.

On Vitiated Air. A Paper read before the Association of Medical Officers of Health.

By C. MEYMOTT TIDY, M.B.,

Lecturer on Chemistry and Professor of Medical Jurisprudence at the London Hospital. 6d.

Cutaneous Medicine, and Diseases of the Skin.
By HENRY S. PURDON, M.D., Edin. F.R.C.S., &c.,
Physician to the Belfast General Hospital, Physician to
the Skin Hospital. 6s.

Skin Diseases : an Inquiry into their Parasitic Origin, Con-
nection with Eye Affections, and a Fungoid Theory of.
By JABEZ HOGG,
Surgeon to the Royal Westminster Ophthalmic Hospital,
President of the Medical Microscopical Society, &c. 2s. 6d.
" Sound teaching will be found in Mr. Hogg's treatise."—*London News.*

By the same Author, in Preparation,

The Ophthalmoscope : a Treatise on its Use in Diseases of
the Eye. 4th edition, 10s. 6d.

The Treatment of Chronic Skin Diseases.
By E. D. MAPOTHER, M.D.,
Professor of Physiology in the Royal College of Surgeons
of Ireland. Price 2s. 6d.

Skin Eruptions : their Real Nature and Rational Treatment,
By Dr. BARR MEADOWS, Physician to the National Institution for Diseases of the Skin, &c. Sixth edition, 2s. 6d.

On the Tonic Treatment of Gout. With Cases.
By JAMES C. DICKINSON, M.R.C.S.,
late of the Medical Staff of H.M.'s Bengal Army, and
formerly Staff Surgeon Crimean Army. 2nd edit., 3s. 6d.
"A thoughtful and practical work."—*Public Opinion.*

By the same Author, 2s.,

Suppressed Gout : its Dangers and Treatment ; with an
Appendix on the Uses of the Vals Waters.

By the same Author, 1s. 6d.

Tropical Debility. A Treatise on the Causes and Treatment
of Debility, produced by prolonged residence in the Tropics.

Also, 1s.

Indian Boils : their Varieties and Treatment without Dis-
figurement.

The Training of the Mind for the Study of Medicine.
A Lecture delivered at St. George's Hospital,
By ROBERT BRUDENELL CARTER, F.R.C.S.,
Professor of Ophthalmic Surgery in the Hospital. 1s.
" A remarkable address."—*The Lancet.*
" No one can read it without learning and profiting much."—*Stud. Jour.*

Osteology for Students, with Atlas of Plates.
By ARTHUR TREHERN NORTON, F.R.C.S.,
Surgeon to, and Lecturer on Anatomy at, St. Mary's Hospital. Atlas and text bound in one volume, price 7s. 6d.,
in two vols. 8s. 6d.
" The handiest and most complete hand-book of Osteology."—*The Lancet.*

In the press by same Author, Second Edition.
Affections of the Throat and Larynx.
" Short, simple, and thoroughly practical instruction.—*Medical Times.*

Lessons in Laryngoscopy : including Rhinoscopy and the
Diagnosis and Treatment of Diseases of the Throat-
Forming a Complete Manual on the Use of the Laryngo.
scope in Diseases of the Throat, Chest, and Lungs.
Illustrated with Hand-coloured Plates and Wood-cuts, for
the use of Practitioners and Students.
By PROSSER JAMES, M.D., M.R.C.P.,
Lecturer on Materia Medica and Therapeutics at the
London Hospital, Physician to the Royal Hospital for
Diseases of the Throat, &c. 5s. 6d.

By the same Author, 1s.,
The Progress of Medicine. A Lecture delivered at the
London Hospital.

Overwork and Premature Mental Decay : its Treatment.
An Essay, with Cases read before the Medical Society of
London.
By C. H. F. ROUTH, M.D., M.R.C.P. London,
Senior Physician to the Samaritan Hospital for Women
and Children, Consulting Physician for Diseases of Women
to the North London Hospital. Price 1s.

By the same Author, price 3s. 6d.,
On Fibrous Tumours of the Womb : Points connected
with their Pathology, Diagnosis, and Treatment. Being
the Lettsomian Lectures delivered before the Medical
Society of London.

Also, by the same Author, price, 1s.
Lectures on Diseases of Women and Children.

Also, price 1s.,
On Uterine Deviations.

The Text-Book of (113) Anatomical Plates, designed under the direction of Professor MASSE, with Descriptive Text.
By E. BELLAMY, F.R.C.S.,
Senior Assistant Surgeon to Charing Cross Hospital.
Second edition, plain, 21*s.*; hand-coloured, 42*s.*
"With these plates the student will be able to read up his anatomy almost as readily as with a recent dissection before him."—*Student's Journal.*

In preparation, price 42s.

The Text Book of Operative Surgery. From the French of Professors CLAUDE BERNARD & HUETTE.
Illustrated with numerous hand-coloured and lithographic plates. Translated and re-written by Arthur Trehern Norton, F.R.C.S., Surgeon to, and Lecturer on Anatomy at, St. Mary's Hospital.

The Students' Case Book: containing Practical Instructions, and all the Necessary Information for Clinical Work and Systematic Case-taking, with a number of Blank Ruled Sheets, for recording full particulars of cases as seen.
By GEORGE BROWN, M.R.C.S.,
Gold Medalist, Charing Cross Hospital. 1*s.*

Elements of the General and Minute Anatomy of Man and the Mammalia. From Original Researches.
By Professor GERBER, University of Bern, and
Professor GEORGE GULLIVER, F.R.S.
2 vols., containing thirty-four plates. 15*s.*

The Pathological Anatomy of the Human Body. Translated from the German of Professor VOGEL.
By GEORGE E. DAY, M.A. Cantab., M.R.C.P. London.
With 100 plain and coloured engravings. 18*s.*

Anatomy of the External Forms of Man: designed for the Use of Artists, Sculptors, &c.
By Dr. J. FAU.
Edited, with Additions, by R. KNOX, M.D., F.R.C.S.E.
Twenty-nine Drawings from Nature. Folio. Plain, 24*s.*; hand-coloured, 42*s.*

Muscular Anatomy of the Horse.
By J. I. LUPTON, M.R.C.V.S. 3*s.* 6*d.*

Cholera : how to Prevent and Resist it.
By Professor VON PETTENKOFER,
Professor of Hygiène, University of Munich, President of
the Sanitary Department of the German Empire ; and
THOMAS WHITESIDE HIME, A.B., M.B.,
Lecturer on Medicine at the Sheffield School of Medicine.
Illustrated with Woodcuts and Diagram, 3*s*. 6*d*.

Diarrhœa and Cholera : their Successful Treatment.
By JOHN CHAPMAN, M.D., M.R.C.P. London. 1*s*. 6*d*.

" His arguments are enforced at great length in the pamphlet before us,
and are supported by an extensive array of facts."—*Medical Times and
Gazette.*

Causes of Cholera : its Treatment.
By WM. GROVE GRADY, M.D., M.R.C.S. 1*s*.

Deafness : its Causes and Treatment, with Anatomy and
Physiology, Human and Comparative, of the Organ of
Hearing ; the Diseases incidental to its Structure, and
their Treatment.
By JOHN P. PENNEFATHER, M.K.Q.C.P., L.R.C.S., &c.,
Surgeon to the Royal Dispensary for Diseases of the
Ear, &c. Illustrated, 5*s*.

Diagnostics of Aural Disease. Second Edition, with a
Chapter on the Application of Electricity, and Description
of the Author's Magneto-Electric Catheter.
By S. E. SMITH, M.R.C.S. Illustrated, 2*s*. 6*d*.

The Philosophy of Voice. An Essay upon the Physiological and Physical Action of the Breath and Vocal Cords
in the Production of Articulate Speech, conjoined with
Vocal Utterance. By CHARLES LUNN. 1*s*.

Experimental Researches on the Causes and Nature of
Hay Fever. With Wood-cuts and Lithographic Tables.
By CHARLES H. BLACKLEY, M.R.C.S. 7*s*.

" It is a piece of real honest work, original and instructive, and will well
repay perusal."—*The Lancet.*

" We have read Mr. Blackley's very instructive treatise with much interest, and have been much impressed by his ingenuity in devising experiments, his industry in carrying them out, and his obvious candour in giving
the results of his observations."—Dr. George Johnson in the *London Medical
Record.*

A Treatise on Pharmacy, designed as a Text-book for Students, and as a Guide for the Physician and Pharmacist, containing the officinal and many unofficinal Formulas, and numerous examples of extemporaneous Prescriptions.
By EDWARD PARRISH,
Late Professor of the Theory and Practice of Pharmacy, Philadelphia College.   Fourth Edition, enlarged and revised, with 280 Illustrations, half-bound morocco, 30*s.*
By THOMAS S. WIEGAND.
" There is nothing to equal Parrish's Pharmacy in this or any other language."—*Pharmaceutical Journal.*

The Pharmacopœial Companion to the Visiting List; Being a Posological Table of all the Medicines of the British Pharmacopœia, arranged according to their action.
By R. T. H. BARTLEY, M.D., M.B. Lond.,
Surgeon to the Bristol Eye Hospital.   Price 6*d.*

The Specific Action of Drugs.   An Index to their Therapeutic Value.   Price 10*s.* 6*d.*
By ALEXANDER G. BURNESS, and F. MAVOR,
President of the Central London Veterinary Society.

The Sewage Question: a Series of Reports.   Being Investigations into the condition of the Principal Sewage Farms and Sewage Works of the Kingdom, from
Dr. LETHEBY's Notes and Chemical Analyses.   4*s.* 6*d.*
"These Reports will dissipate obscurity, and, by placing the subject in a proper light, will enable local authorities, and others interested in the matter, to perceive the actual truths of the question, and to apply them practically."

Notes on Nuisances, Drains, and Dwellings: their Detection and Cure.   By W. H. PENNING, F.G.S.   6*d.*
"The directions, which are plain, sound, and practical, will be found useful in every household."—*The Doctor.*
"This little pamphlet should be studied by everybody."—*Scientific Review.*

On Scarlatina: its Nature and Treatment.
By I. BAKER BROWN, F.R.C.S. (Exam.),
late Surgeon Accoucheur to, and Lecturer on Diseases of Women and Children at, St. Mary's Hospital.   3rd edit. 3*s.*
By the same Author, 1*s.,*
Sterility: its Causes and Treatment.   Being a Paper read before the Medical Society of London, and printed by request.

On Change of Climate in the Treatment of Chronic Diseases, especially Consumption: A Medical Guide for Travellers in pursuit of Health to the Southern Winter Resorts of Europe and Africa, the South of France, Spain, Portugal, Italy, Algeria, the Mediterranean, Egypt, &c. By THOMAS MORE MADDEN, M.D., M.R.I.A., Examiner in Obstetric Medicine in the Queen's University, Ex-Physician to the Rotundo Lying-in-Hospital, &c.

Third Edition, 5s.

"Evidently the work of a well-informed physician."—*The Lancet.*
"Such a book is very opportune."—*The Athenæum.*

By the same Author. Third Edition, 5s.,

The Spas and their Use. A Medical Handbook of the Principal Watering Places on the Continent resorted to in the Treatment of Chronic Diseases, especially Gout, Rheumatism, and Dyspepsia, with Notices of Spa Life, and Incidents of Travel.

"Not only full of matter, but withal most readable, chatty, and interesting."—*British Medical Journal.*
"A useful handbook for both the professional and the general reader."—*The Lancet.*

By the same Author, Second Edition, Price 1s.,

The Diseases of Women connected with Chronic Inflammation of the Uterus. Their constitutional character and treatment.

By the same Author, Demy 8vo., price 1s.,

The Diagnosis and Treatment of Uterine Polypi.

Also, Royal 8vo., price 1s.,

On Uterine Hydatidiform Disease, or Cystic Degeneration of the Ovum.

Typhoid Fever: its Treatment. By WILLIAM BAYES, M.D., L.R.C.P. 1s.

African, West Indian, and other Fevers and Diseases. By ALEXANDER LANE, M.D., Surgeon Royal Navy. 6d.

How to Prevent Small Pox. By MORDEY DOUGLAS, M.R.C.S., L.R.C.P. Third edition 6d.

"This is a very valuable pamphlet."—*Medical Press.*

Short Lectures on Experimental Chemistry.   Introductory to the general course.
   By J. EMERSON REYNOLDS, F.C.S., M.R.C.P., Professor of Chemistry, Royal College of Surgeons, Professor of Analytical Chemistry, and Keeper of the Minerals, Royal Dublin Society. 3s. 6d.

Notes on the Pharmacopœial Preparations.   Specially arranged for the use of Students preparing for Examinations, and as a Note-book for General Practitioners.
   By HANDSEL GRIFFITHS, Ph.D., L.R.C.P., &c., Librarian to the Royal College of Surgeons of Ireland. 2s. 6d.

"Will be found useful to students engaged in the study of Materia Medica and the Pharmacopœia; the Notes are faithful."—*The Lancet.*

"From the many excellences of the work, we can confidently recommend it as a most valuable help for those who are preparing for medical examinations."—*Students' Journal.*

By the same Author, third edition, 1s. 6d.,

Posological Tables: Being a Classified Chart of Doses; showing at a glance the Dose of every Officinal Substance and Preparation. For the use of Practitioners and Students.

"We welcome these Tables, which are the best we have seen, as a great boon to students and practitioners."—*Hospital Gazette.*

"The Local Government Board might advantageously consider the propriety of supplying every dispensary and workhouse with a copy, the cost of which would weigh little against the benefit which such ready information would be to the Poor Law service."—*Medical Press and Circular.*

Also by the same Author, 1s. 6d.,

A System of Botanical Analysis, applied to the Diagnosis of British Natural Orders, for the Use of Beginners.

"Backed by such high authority as Professors Bentley, Henslow, and other eminent botanists, we can safely introduce it to the notice of our readers."—*Students' Journal.*

"The author has placed the student under considerable obligations by his system of botanica analysis."—*Pharmaceutical Journal.*

Practical Observations on the Harrogate Mineral Waters.
   By A. S. MYRTLE, M.D., L.R.C.S.E.
   Third Edition, 2s. 6d.

Chemistry in its Application to the Arts and Manufactures. A Text-Book by RICHARDSON and WATTS.

Vol. I. : Parts 1 and 2.—Fuel and its Applications. 433 Engravings,
and 4 Plates . . . . £1 16s.

Part 3.—Acids, Alkalies, Salts, Soap, Soda, Chlorine and its
Bleaching Compounds, Iodine, Bromine, Alkalimetry,
Glycerine, Railway Grease, &c., their Manufacture
and Applications . . . £1 13s.

Part 4.—Phosphorus, Mineral Waters, Gunpowder, Gun-
cotton, Fireworks, Aluminium, Stannates, Tung-
states, Chromates and Silicates of Potash and Soda,
Lucifer Matches . . . £1 1s.

Part 5.—Prussiate of Potash, Oxalic Acid, Tartaric Acid,
Many Tables, Plates, and Wood Engravings, £1 16s.

Parts 3, 4, and 5 separately, forming a complete

Practical Treatise on Acids, Alkalies, and Salts : their Manufacture and Application. In three vols., £4 10s.

Collenette's Chemical Tables : Oxides, Sulphides, and Chlorides, with Blank Forms for Adaptation to other Compounds. Arranged for the use of Teachers and Students, by Professor COLLENETTE. 6d.

" We have great pleasure in recommending this little work to all who are interested in having the study of chemistry simplified and methodically treated."—*Chemical News.*

" An excellent means for the communication of much valuable information."—*Chemist and Druggist.*

Chemistry in its Relation to Physiology and Medicine. By GEORGE E. DAY, M.A. Cantab., M.D., F.R.S., late Professor of Medicine in the University of St. Andrews. 10s.

A Practical Text-Book of Inorganic Chemistry, including the Preparation of Substances, and their Qualitative and Quantitative Analyses, with Organic Analyses. By D. CAMPBELL, late Demonstrator of Practical Chemistry in University College. 5s. 6d.

Rudiments of Chemistry, with Illustrations of the Chemistry of Daily Life. Fourth Edition, with 130 Woodcuts. By D. B. REID, M.D., F.R.S., F.R.C.P. Edin. 2s. 6d.

The Chemical and Physiological Balance of Organic Nature : an Essay. 1 vol., 12mo. By Professors DUMAS and BOUSSINGAULT. 4s.

Elements of Chemistry; including the application of the Science in the Arts.
By T. GRAHAM, F.R.S.,
late Master of the Mint. Second Edition, revised and enlarged. Illustrated with Woodcuts. 2 vols., 8vo. £2. Vol. II. Edited by H. WATTS, M.C.S. Separately, £1.

Practical Treatise on the Use of the Microscope.
By J. QUECKETT.
Illustrated with 11 Steel Plates and 300 Wood Engravings. Third Edition, £1 1s.

Lectures on Histology: Elementary Tissues of Plants and Animals. On the Structure of the Skeletons of Plants and Invertebrate Animals. 2 vols., 8vo. Illustrated by 340 Woodcuts. By the same Author. £1 8s. 6d.

Introduction to Cryptogamic Botany. 8vo. Illustrated with 127 Engravings. By Rev. M. J. BERKELEY. £1.

A Practical Treatise on Coal, Petroleum, and other Distilled Oils. Illustrated with 42 Figures, and a View on Oil Creek, in Pennsylvania. 8vo.
By A. GESNER. 10s. 6d.

Practical Mineralogy; or, a Compendium of the Distinguishing Characters of Minerals, by which the Name of any Species may be speedily ascertained. 8vo., with 13 Engravings, showing 270 Specimens.
By E. J. CHAPMAN. 7s.

Schleiden's Plants: a Biography, in a Series of Fourteen Popular Lectures on Botany.
Edited by Professor HENFREY.
Second Edition, 8vo., with 7 Coloured Plates and 16 Woodcuts. 15s.

The Architecture of the Heavens.
By J. P. NICHOL,
Professor of Astronomy in the University of Glasgow Ninth Edition, entirely revised and greatly enlarged. Illustrated with 23 Steel Engravings and numerous Woodcuts. 16s.

The Protoplasmic Theory of Life. Containing the Latest Researches on the subject.

By JOHN DRYSDALE, M.D., F.R.M.S.,

President of the Liverpool Microscopical Society. 5*s.*

"Subjects beyond the pale of precise knowledge are treated of in a manner which will quite repay perusal."—*Nature.*

By the same Author, |

Life and the Equivalence of Force.

Part I. Historical Notice of the Discovery of the Law of Equivalence of Force. 1*s.*

Part II. Nature of Force and Life : containing the Harmony of Fletcher and Beale. 1*s. 6d.*

"The book is well worth perusal."—*Westminster Review.*

"We cannot part from this work without praising the calm and excellent spirit in which the subject is handled."—*The Examiner.*

Practical Lessons in the Nature and Treatment of the Affections produced by the Contagious Diseases; with Chapters on Syphilitic Inoculation, Infantile Syphilis, and the Results of the Contagious Diseases Acts. Sixty coloured and plain Illustrations.

By JOHN MORGAN, M.D., F.R.C.S.,

Professor of Anatomy in the Royal College of Surgeons, Lecturer on Clinical Surgery, Physician to the Lock Hospitals, Dublin. Second thousand. Paper wrapper, 5*s.* ; cloth, 6*s.*

"Contains much that is original and of practical importance."—*The Lancet.*

"This is a most instructive work, and reflects great credit on Dr. Morgan."—*The Medical Press and Circular.*

By the same Author. Illustrated, price 1s.

On the Cure of Bent Knee, and the immediate Treatment of Stiff-joints by extension.

Also, Second Thousand, Price 2s.

The Dangers of Chloroform and the Safety and Efficiency of Ether as an Agent in securing the Avoidance of Pain in Surgical Operations.

The Dental Profession. A Letter by a Dental Surgeon, Member of the Royal College of Surgeons, on the abuse of the profession. Price 1*s.*

Responsibility and Disease : an Essay upon moot-points in Medical Jurisprudence, about which Medical Men should not fail to be well instructed.

By J. H. BALFOUR BROWNE, Barrister-at-Law,
Author of " The Medical Jurisprudence of Insanity," &c. *2s.*

Diseases of the Prostrate Gland.

By J. STANNUS HUGHES, M.D., F.R.C.S.,
Professor of Surgery, Royal College of Surgeons, Vice-President of the Dublin Pathological Society. Revised Edition, *3s.*

Syphilis : Its Nature and Treatment.

By CHARLES R. DRYSDALE, M.D., F.R.C.S.,
Physician to the Metropolitan Free Hospital ; late Secretary Harveian Medical Society's Committee for the Prevention of Venereal Diseases. Second Edition, *4s. 6d.*

"We bespeak a cordial welcome to this new work, which contains in a moderate compass the conclusions of an industrious, painstaking syphilographer."—*Medical Press and Circular.*

By the same Author,

Alpine Heights and Climate in Consumption. *1s.*

Medicine as a Profession for Women. *1s.*

The Population Difficulty. *6d.*

Functional Derangements and Debilities of the Generative System : their Nature and Treatment.
By F. B. COURTENAY, M.R.C.S. Eighth Edition, *3s.*

On Certain Forms of Hypochondriasis, and Debilities peculiar to Man. Translated from the German of DR. PICKFORD,
By F. B. COURTENAY, M.R.C.S.
Eighth Edition, price *5s.*

Modern Hydropathy : with Practical Remarks upon Baths, in Acute and Chronic Diseases.
By JAMES WILLIAMS, M.D., M.R.C.S. Fifth Edition. limp cloth, *2s.*

Practical Guide to the Baths of Aix in Savoy.
By the Baron DESPINE, Physician. *2s.*

Horses : their Rational Treatment, and the Causes of their Premature Decay. By Amateur. *5s.*

An Abridgment of the above. *1s.*

Via Medica : a Treatise on the Laws and Customs of the Medical Profession, in relation especially to Principals and Assistants ; with Suggestions and Advice to Pupils on Preliminary Education.

~ By J. BAXTER LANGLEY, LL.D., M.R.C.S., F.L.S. Fourth Edition, 3*s.*

Engineering Precedents for Steam Machinery : embracing the Performances of Steamships, Experiments with Propelling Instruments, Condensers, Boilers, &c., accompanied by Analyses of the same ; the whole being original matter, and arranged in the most practical and useful manner for Engineers. 2 vols., 8vo. With Plates and Tables.

By B. E. ISHERWOOD, Chief Engineer United States Navy. 15*s.*

Dictionary of Technical Terms used in Iron Ship-building, Steam-engines, &c. In English, French, and Latin.

By GIORGIO TABERNA. 3*s.*

On Mental Capacity in Relation to Insanity, Crime, and Modern Society.

By CHRISTOPHER SMITH, M.D. Price 3*s.* 6*d.*

WORKS BY DR. ROTH.

1. **The Prevention and Cure of many Chronic Diseases** by Movements. With Ninety Engravings. 10*s.*

2. **The Handbook of the Movement-Cure.** With One Hundred and Fifty-five Original Engravings. 10*s.*

3. **Contributions to the Hygienic Treatment of Paralysis, and of Paralytic Deformities.** With Thirty-Eight Engravings. Illustrated by Numerous Cases. 3*s.* 6*d.*

4. **The Prevention of Spinal Deformities,** especially of Lateral Curvature ; with Notes. 3*s.* 6*d.*

5. **On Paralysis in Infancy, Childhood, and Youth.** With Forty-five Engravings. 3*s.* 6*d.*

6. **A Short Sketch of Rational Medical Gymnastics,** or the Movement-Cure. With Thirty-eight Engravings. 1*s.*

7. **Table, showing a few Injurious Positions,** and some Deformities of the Spine, produced partly by bad positions, and tight lacing. With Forty-six Engravings. 6*d.*

8. **A Table of a Few Gymnastic Exercises without** Apparatus. With Thirty-three Wood Engravings. 6*d.*

9. **Gymnastic Exercises on Apparatus,** according to the Rational System of LING. With Eighty Illustrations. 1*s.*

10. **The Gymnastic Exercises of Ling.** Arranged by Dr. Rothstein ; translated by Dr. Roth. Second Edition. 2*s.* 6*d.*

11. **The Russian Bath :** with some Suggestions regarding Public Health. Second Edition. 1*s.*

12. **On the Causes of the Great Mortality of Children,** and the Means of Diminishing them. 3*d.*

13. **On the Importance of Rational Gymnastics** as a Branch of National Education : a Letter to Lord Granville. 1*s.*

14. **On Scientific Physical Training and Rational Gymnastics ;** a Lecture. 1*s.*

15. **Exercises or Movements,** according to LING's System: With Forty-two Illustrations. Fourth Edition. 1*s.*

16. **Two Tables of Gymnastic Exercises without Apparatus ;** with Explanations. 1*s.*

17. **A Plea for the Compulsory Teaching of Physical Education.** 1*s.*

Practical Guide for the Young Mother. Translated
from the French of Dr. BROCHARD, late Director-General
of Nurseries and Crèches in France, Edited with Notes
and Hints for the English Mother.

Crown 8vo., 2s.

TABLE OF CONTENTS.

FIRST PART.

PHYSICAL EDUCATION OF THE INFANT.

SECOND PART.

MISCELLANEOUS.

THE CRY OF ITALY AGAINST THE ROMISH CHURCH.

The Religion of Rome described by a Roman.   8s.

Translated, with Introduction and Notes,

By WILLIAM HOWITT,

Author of "The History of Priestcraft," "Homes and Haunts of British Poets," &c., &c.

"Should be read by all."—*Standard.*

"The object is to prove the emptiness of Roman Ecclesiastical Religion, and the moral unfitness of many of its chief professors to act as religious teachers.   A closing chapter upon Catholicism in Spain, and the influence of the priesthood on political affairs in that country, is of special interest at the present moment."—*The Mail* (*Evening Edition of The Times*).

"Mr. Howitt has seen Old Giant Pope at home, and marked for himself the monster's baleful influence.   To his testimony we can add our own corroborating witness, and so, we believe, can every sojourner in Italy. Written with great vigour and vivacity."—*Mr. Spurgeon in the "Sword and Trowel."*

"The value of the work lies in its contemporary character."—*Literary Churchman.*

"A scathing and unflinching revelation of the iniquities inseparable from the Papal system."—*The Rock.*

"This is a book which we hope will excite much public attention."—*Evangelical Magazine.*

"The book bristles with facts which ought to startle and arouse."—*Surrey Congregational Magazine.*

"For a real view of the blessings of Popery we must study its growth, and development, and influence in its own real home.   The book before us enables us to understand something of what it is and what it has done."—*Literary World.*

"There is no probability of controverting such a work as this, and if it were generally read throughout England, we believe it would give Popery a deadly wound in this country."—*Protestant Opinion.*

"This is in every sense a seasonable book, and deserves a wide circulation."—*Watchman.*

"The book remains an exhaustive and telling indictment against the whole pontifical system."—*The Examiner.*

Cheerful Words: Volumes of Sermons, specially adapted for delivery before Inmates of Asylums, Unions, Workhouses, Hospitals, Gaols, Penitentiaries, and other Public Institutions. Composed by distinguished Dignitaries of the Church, and Clergymen. Edited by WM. HYSLOP, Proprietor of the Stretton House Private Asylum for Gentlemen, Church Stretton, Shropshire. First and Second Series, price 5*s.* each.

Hymnologia Christiana Latina; or, a Century of Psalms and Hymns and Spiritual Songs. By Various Authors, from LUTHER to HEBER. Translated into Latin Verse by the Rev. RICHARD BINGHAM, M.A. 5*s.*

"There are something under a hundred and twenty versions in this elegant little volume, many of them of considerable length. It is valuable, merely as an evidence of that elaborate cultivation, that perfect polish of classical scholarship, which it is very good for the world at large that some men should possess."—*Literary Churchman.*

A Physician's Sermon to Young Men. By WILLIAM PRATT, M.A., M.D., &c. 1*s.*

"The delicate topic is handled wisely, judiciously, and religiously, as well as very plainly."—*Guardian.*

Electricity Made Plain and Useful. By JOHN WESLEY, M.A. Second Edition, 2*s.* 6*d.* A Popular Edition, 1*s.*

"A curious and entertaining little work."—*Literary Churchman.*

Glimpses of a Brighter Land. Cloth extra, 2*s.* 6*d.*

Manual Alphabet for the Deaf and Dumb. Official. 6*d.*

Brilliant Prospects. A Novel. By R. L. JOHNSON, M.D. 3*s.*6*d.*

Queer Customers. By the same Author. 1*s.*

My First Start in Practice. By the same Author. 1*s.*

Constipation: its Causes and Consequences. With Hints. By a CLERGYMAN. 1*s.*

Patent Wrinkles. With Practical Suggestions, written in a humorous style, for Amendment of the Patent Laws. 1*s.*

Revelations of Quacks and Quackery. Giving a complete Directory of the London and Provincial Quack Doctors; with Facts and Cases in Illustration of their Nefarious Practices. Twenty-fifth thousand. 1*s.* 6*d.*

"The narrative is too good to be abridged, and ought to be, as we believe it is, largely circulated, which is no less than it deserves, both for its fearless tone, and for the care and research which have been bestowed on its compilation."—*Saturday Review.*

"Buy, therefore, reader, by all means buy 'Revelations of Quacks and Quackery.' Its contents will amuse and astonish you, while they invoke your indignation and disgust."—*Punch.*

PERIODICAL PUBLICATIONS.

The Medical Press and Circular. Established 1838. Published every Wednesday Morning in London, Dublin, and Edinburgh. Is one of the oldest and most influential of the Medical Journals. 5*d*. Per annum, post free, in advance, £1 1*s*.

The Student's Journal and Hospital Gazette. A Fortnightly Review of Medicine, Surgery, Arts, Science, Literature, and the Drama. The only Paper that represents the whole body of Medical Students in the United Kingdom. 4*d*. Per annum, post free, 7*s*. 6*d*.

The Doctor. A Monthly Review of British and Foreign Medical Practice and Literature. Published on the 1st of every Month. 6*d*. Per annum, post free, 6*s*.

Anthropologia. The Quarterly Journal of the London Anthropological Society. 4*s*. each part.

The Ecclesiastical Gazette ; or, Monthly Register of the Affairs of the Church of England. Established 1838. Published on the Second Tuesday in every Month, and sent to the Dignitaries of the Church at home and abroad, Heads of Colleges, and the Clergymen of every Parish in England and Wales. 6*d*. Per annum, post free, 6*s*.

The Clergy List (Annual). Established 1841. Contains Alphabetical Lists of the Clergy at Home and Abroad. Benefices in England and Wales, with Post Towns, Incumbents, Curates, Patrons, Annual Value, Population, &c. The Patronage of the Crown, Lord Chancellor, Archbishops, Bishops, Deans, Universities, Private Patronage, &c., &c. 10*s*.

The Irish Medical Directory (Annual). Contains a complete Directory of the Profession in Ireland ; their Residences and Qualifications ; the Public Offices which they hold, or have held ; the Dates of Appointments ; and the published Writings for which they are distinguished. 5*s*.

The Medical Register and Directory of the United States of America. Containing the Names and Addresses of about 70,000 Practitioners of all grades, systematically arranged by States. 30*s*.

STANDARD FRENCH WORKS.

	£	s.	d.
Alvarenga.—Thermométrie clinique	0	5	0
Anger.—Nouveaux éléments d'anatomie chirurgicale, avec atlas	2	0	0
—— Maladies chirurgicales nouvelles fractures et luxations, coloriées	7	10	0
Anglada.—Etudes sur les maladies nouvelles et les maladies éteintes	0	8	0
Armand.—Traité de Climatologie générale	0	14	0
Barnes.—Leçons sur les opérations obstétricales	0	12	0
Barthes et Rilliet.—Traité clinique et pratique des maladies des enfants — 3 *vols.*	1	5	0
Bayard.—Traité pratique des maladies de l'estomac	0	10	0
Beaude.—Dictionnaire de médecine usuelle à l'usage des gens du monde — 2 *vols.*	1	10	0
Beaunis et Bouchard.—Nouveaux éléments d'anatomie descriptive	0	18	0
Becquerel.—Traité des applications de l'électricité à la thérapeutique médicale et chirurgicale	0	7	0
—— Traité élémentaire d'hygiène privée et publique	0	8	0
Beraud.—Atlas complet d'anatomie chirurgicale topographique	3	0	0
—— Ditto ditto, with coloured plates	6	0	0
—— et Robin.—Manuel de physiologie de l'homme et des principaux vertébrés — 2 *vols.*	0	12	0
—— et Velpeau.—Manuel d'anatomie générale et thérapeutique	0	7	0
Bergeret.—Abus des boissons alcooliques	0	3	0
—— Fraudes dans l'accomplissement des fonctions génératrices	0	2	6
Bernard-Chevell.—Leçons de physiologie expérimentale appliquée à la médecine	0	14	0
—— Leçons sur les effets des substances toxiques et médicamenteuses	0	7	0
—— Leçons sur la physiologie et la pathologie du système nerveux — 2 *vols.*	0	14	0
—— Leçons sur les propriétés physiologiques et les altérations pathologiques des liquides de l'organisme 2 *vols.*	0	14	0
—— Leçons de pathologie expérimentale	0	7	0
—— De la physiologie générale	0	6	0
—— et Huette.—Précis iconographique de médecine opératoire et d'anatomie chirurgicale	1	4	0
—— Ditto ditto, with coloured plates	2	8	0
—— Ditto ditto, plain	0	3	0
—— Ditto ditto, coloured	0	6	0
—— Premiers secours aux blessés	0	2	0
Best.—Leçons sur la physiologie comparée de la respiration	0	10	0

	£	s.	d.
Billroth.—Eléments de Pathologie Chirurgicale général -	o	14	o
Bocquillon.—Manuel d'histoire naturelle médicale - *2 vols.*	o	14	o
Boisseau.—Des maladies simulées et des moyens de les reconnaître -	o	7	o
Boivin et Duges.—Anatomie pathologique de l'uterus et de ses annexes	2	5	o
Bonnafont.—Traité théorique et pratique des maladies de l'oreille, et des organes de l'audition -	o	10	o
Bouchardat.—Le Travail, son influence sur la santé -	o	2	6
———— Annuaire de thérapeutique, de matière médicale, de pharmacie, et de toxicologie -	o	1	3
———— Formulaire vétérinaire	o	4	6
———— Manuel de matière médicale, de thérapeutique et de pharmacie *2 vols.*	o	16	o
———— Nouveau formulaire magistral -	o	3	6
Bouchut.—Histoire de la médecine et des doctrines Médicales *2 vols.*	o	16	o
———— Traité de pathologie générale et de séméiotique	1	o	o
———— De la vie et de ses attributs -	o	3	6
———— Traité pratique des maladies des nouveau-nés -	o	16	o
———— et Despres.—Dictionnaire de médecine et de thérapeutique	1	5	o
Boudin.—Traité de géographie et de statistique médicales, et des maladies endémiques *2 vols.*	1	o	o
Bourgery.—Traité de l'anatomie de l'homme, comprenant la médecine opératoire, dessiné d'après nature, par H. Jacob—8 vols. folio, with 726 plates -	30	o	o
———— Ditto ditto, with coloured plates	50	o	o
———— et Jacob.—Anatomie élémentaire en 20 planches, représentant chacune un sujet dans son entier à la proportion de demi-nature, avec un texte explicatif	10	o	o
———— Ditto ditto, coloured -	20	o	o
Bourgeois.—Les passions dans leurs rapports avec la santé et les maladies	o	2	o
Brehm.—La vie des animaux illustrée—Les mammifères *2 vols.*	1	1	o
———— Ditto ditto Les oiseaux *2 vols.*	1	1	o
Briand et Chaude.—Manuel complet de médecine légale -	o	18	o
Burdel.—Du cancer considéré comme souche tuberculeuse -	o	3	o
Carles.—Etude sur les quinquinas	o	2	6
Casper.—Traité pratique de médecine légale, traduit de l'Allemand par M. G. Baillière *2 vols.*	o	15	o
———— Coloured atlas separately -	o	12	o
Cauvet.—Nouveaux éléments d'histoire naturelle médicale, *2 vols.*	o	12	o
Cerise.—Mélanges médico-psycologiques -	o	7	6
Chailly-Honore.—Traité pratique de l'art des accouchements -	o	10	o

	£	s.	d.
Chauffard.—De la fièvre traumatique, etc.	0	3	6
Chauveau.—Traité d'anatomie comparée des animaux domestiques	1	0	0
Civiale.—Traité pratique sur les maladies des organes génito-urinaires - - - 3 *vols.*	1	4	0
Codex Médicamentarius, pharmacopée française, rédigée par ordre du gouvernement	0	10	0
Colin.—Traité de physiologie comparée des animaux 2 *vols.*	1	6	0
Comite-Consultatif.—d'hygiène publique de France, recueil des travaux et des actes officiels de l'administration sanitaire, chaque vol.	0	8	0
Comte.—Structure et physiologie de l'homme, demontrées à l'aide des figures coloriées, découpées, et superposées	0	4	6
Corlieu.—Aide-mémoire de médecine, de chirurgie, et d'accouchements	0	6	0
Cornil et Ranvier.—Manuel d'histologie pathologique. Parts I. and II., chaque	0	4	6
Coze et Feltz.—Recherches cliniques et expérimentales sur les maladies infectieuses	0	6	0
Cruveilhier. — Traité d'anatomie pathologique générale 5 *vols.*	1	15	0
———— Anatomie pathologique du corps humain. 41 livraisons, chaque	0	11	0
Cuvier.—Les oiseaux décrits et figurés. 72 planches, 464 figures, noires	1	10	0
———— Ditto ditto, coloriées	2	10	0
———— Les mollusques. 56 planches, 520 figures, noires	0	15	0
———— Ditto ditto, coloriées	1	5	0
———— Les vers et les zoophytes. 37 planches, 520 figures, noires	0	15	0
———— Ditto ditto coloriées	1	5	0
Cyon.—Principes d'électrothérapie	0	4	0
Cyr.—Traité d'alimentation, dans ses rapports avec la physiologie, la pathologie, et la thérapeutique	0	8	0
Daremberg.—Histoire des sciences médicales 2 *vols.*	1	0	0
———— Médecine, histoire et doctrines	0	3	6
Davaine.—Traité des entozoaires et des maladies vermineuses de l'homme et des animaux domestiques	0	12	0
Demarquay.—De la régénération des organes et des tissus	0	16	0
Deschampes.—Compendium de pharmacie pratique	1	0	0
Desmarres.—Chirurgie oculaire	0	8	0
Despres.—Rapport sur les travaux de la 7ème ambulance à l'armée du Rhin et à l'armée de la Loire	2	0	0
Dolbeau.—Leçons de clinique chirurgicale	0	7	0
———— De la lithotritie périnéale	0	4	0
Donne.—Hygiène des gens du monde	0	4	0
Dorvault.—Officine ou répertoire générale de pharmacie pratique	0	17	0

	£	s.	d.
Duchartre.—Eléments de botanique, comprenant l'anatomie, l'organographie, la physiologie des plantes, les familles naturelles, et la géographie botanique -	o	18	o
Duchenne.—De l'électrisation localisée - - -	o	18	o
Durand-Fardel.—Traité pratique des maladies chroniques *2 vols.*	I	o	o
————— Dictionnaire général des eaux minérales et d'hydrologie médicale - - - *2 vols.*	I	o	o
————— Traité pratiques des maladies des vieillards -	o	14	o
————— Traité clinique et thérapeutique du diabète -	o	5	o
Duval et Lerebouilet.—Manuel du microscope -	o	5	o
Farabeuf.—De L'épiderme et des épithéliums - -	o	5	o
Follin et Duplay.—Traité élémentaire de pathologie externe. Vols. I., II., et III. - - -	I	17	o
Fonssagrives.—Hygiène et assainissement des Villes -	o	8	o
Fort.—Pathologie et clinique chirurgicales - *2 vols.*	I	5	o
Foville.—Etude clinique de la folie, avec prédominance du délire des grandeurs - - -	o	4	o
————— Moyens de combattre l'ivrognerie - -	o	5	o
Galante.—Emploi du caoutchouc vulcanisé dans la thérapeutique médico-chirurgicale - - -	o	5	o
Galezowski.—Traité des maladies des yeux - -	I	o	o
————— Du diagnostic des maladies des yeux par la chromatoscopie rétinienne - - .	o	7	o
Gallard.—Leçons cliniques sur les maladies des femmes -	o	12	o
Gallez.—Histoire des kystes de l'ovaire - -	o	12	o
Ganot.—Traité élémentaire de physique, expérimentale et appliquée, et de météorologie - -	o	7	o
Garnier.—Dictionnaire annuel du progrès des sciences et institutions médicales - - -	o	7	o
Garrigou.—Bagnères de Luchon - - -	o	8	o
Gaujot et Spillmann.—Arsenal de la chirurgie contemporaine - - - *2 vols.*	I	12	o
Gervais et Van Benenden.—Zoologie médicale -	o	15	o
Gintrac.—Cours théorique et clinique de pathologie interne et de thérapie médicale - - *9 vols.*	3	3	o
Girard.—Traité élémentaire d'entomologie coléoptères. Avec atlas, colorié - - - -	3	o	o
————— Ditto, ditto, noire - - - - -	I	10	o
Gloner.—Nouveau dictionnaire de thérapeutique - -	o	7	o
Godron.—De l'espèce et des races dans les êtres organisées -	o	12	o
Goffres.—Précis iconographique des bandages, pansements, et appareils - - - - -	I	16	o
————— Ditto, ditto, in parts, plain - - -	o	3	o
————— Ditto, ditto, in parts, coloured -	o	6	o
Gori.—Des hôpitaux, tentes, et baraques - -	o	3	o
Gosselin.—Clinique chirurgicale de l'hôpital de la Charité *2vols*	I	4	o
Goubert.—Manuel de l'art des autopsies cadavériques, surtout dans les applications à l'anatomie pathologique	o	6	o

	£	s.	d.
Graefe.—Clinique ophthalmique	o	8	o
Grehant.—Manuel de physique médicale	o	7	o
Grellois.—Histoire médicale du blocus de Metz	o	6	o
Gubler.—Commentaires thérapeutiques du codex médicamentarius	o	13	o
Guibourt.—Histoire naturelle des drogues simples 4 *vols.*	1	16	o
Guyon.—Eléments de chirurgie clinique	o	12	o
Hacquart.—Botanique médicale	o	6	o
Herard et Cornil.—De la phthisie pulmonaire	o	10	o
Houel.—Manuel d'anatomie pathologique générale	o	7	o
Jamain.—Manuel de pathologie et de clinique chirurgicales 2 *vols.*	o	15	o
—— Manuel de petite chirurgie	o	7	o
—— Nouveau traité élémentaire d'anatomie descriptive et des préparations anatomiques	o	12	o
—— Figures coloriées	2	o	o
Jeannel.—Prostitution dans les grandes villes aux dix-neuvième siècle	o	4	
—— Formulaire magistral et officinal international	o	6	o
Jobert (de Lomballe).—De la réunion en chirurgie	o	12	o
Kiess et Duval.—Cours de physiologie	o	7	o
Kiener.—Le Spécies général et iconographie des coquilles vivantes, continué par le Docteur Fischer. Genre Turbo, avec 43 planches gravies et coloriées	2	10	o
—— Genre Trochus (paraîtra prochainement)			
Lancereaux.—Atlas d'anatomie pathologique	4	o	o
Lecour.—Prostitution à Paris et à Londres	o	4	6
Le Fort.—La chirurgie militaire et les sociétés de secours en France et à l'étranger	o	10	o
Lefort.—Traité de chimie hydrologique	o	12	o
Legouest.—Traité de chirurgie de l'armée	o	14	o
Lemaire.—Acide phénique	o	6	o
Levy.—Traité d'hygiène publique et privée 2 *vols.*	1	o	o
Liebriech.—Atlas d'ophthalmoscopie représentant l'état normal et les modifications pathologiques du fond de l'œil visibles à l'ophthalmoscope	1	10	o
Littre et Robin.—Dictionnaire de médecine, de chirurgie, de pharmacie, de l'art vétérinaire et des sciences qui s'y rapportent 2 *vols.*	1	o	o
Longet.—Traité de Physiologie, 3ème edition	1	16	o
Lorain.—Etudes de médecine clinique : Le choléra, observé à l'hôpital Saint Antoine	o	7	o
—— Le pouls, ses variations et ses formes diverses dans les maladies	o	10	o
Luys.—Iconographie photographique des centres nerveux	7	10	o
Mailliot.—Auscultation	o	12	o
Malgaigne.—Manuel de médecine opératoire	o	7	o
—— Traité d'anatomie chirurgicale et de chirurgie expérimentale 2 *vols.*	o	18	o

		£	s.	d.
Mandl.—Maladies du larynx et du pharynx -	•	0	18	0
Marais.—Guide pratique pour l'analyse des urines -	-	0	3	6
Marce.—Traité pratique des maladies mentales	-	0	8	0
——— Recherches cliniques et anatomo-pathologiques	•	0	1	6
Marchant.—Étude sur les maladies épidémiques	-	0	1	0
Marvaud.—Effets physiologiques et thérapeutiques des aliments d'épargne ou antidéperditeurs	•	0	3	6
——— Les Aliments d'épargne Alcool et Boisons aromatiques (café, thé, &c.)		0	6	0
Maunory et Salmon.—Manuel de l'art des accouchements		0	7	0
Mayer.—Rapports conjugaux, considérés sous point de vue de la population, santé, et de la morale publique -		0	3	0
Mayer.—Mémoire sur le mouvement organique dans ses rapports avec la nutrition -	-	0	3	0
Meyer.—Traité des maladies des yeux -	-	0	10	0
Montmeja.—Pathologie iconographique du fond de l'œil, traité d'ophthalmoscope -	•	0	18	0
Moquin-Tandon.—Éléments de botanique médicale	-	0	6	0
——— Éléments de zoologie médicale -	•	0	6	0
Morel.—Traité d'histologie humaine -	-	0	12	0
Naegele et Grenser.—Traité de l'art des accouchements	-	0	12	0
Naquet.—Principes de chimie fondée sur les théories modernes 2 *vols.*		0	10	0
Nelaton.—Éléments de pathologie chirurgicale - 3 *vols.*		1	9	0
Nielly.—Manuel d'obstétrique ou aide-memoire de l'élève et du praticien -	•	0	4	0
Niemeyer.—Pathologie interne - 2 *vols.*		0	14	0
Onimus et Legros.—Traité d'électricité médicale -	-	0	12	0
Penard.—Guide pratique de l'accoucheur et de la sage-femme		0	4	0
Peter.—Leçons de clinique médicale - *vol.* 1		0	15	0
Petrequin.—Nouveaux mélanges de chirurgie et de médecine -	•	0	7	6
Pidoux.—Etudes sur la phthisie -	•	0	9	0
Poggiale.—Traité d'analyse clinique -	-	0	9	0
Poincare.—Lecons sur la Physiologie normale et pathologique du Système Nerveux -		0	4	0
Quatrefages et Hamy.—Les crânes des races humaines, par livraison chaque -	-	0	14	0
Quetelet.—Anthromopetrie, ou mesure des différentes facultés de l'homme -	•	0	12	0
——— Physique sociale, ou essai sur le développement des facultés de l'homme - 2 *vols.*		1	0	0
Raciborski.—Histoire des découvertes relatives au système veineux -	-	0	3	0
——— Traité de la menstruation -	-	0	12	0
Racle.—Traité de diagnostic médical, guide clinique pour l'étude des signes caractéristiques des maladies -		0	7	0
Reindfleisch.—Traité d'histologie pathologique -	•	0	14	0
Requin.—Éléments de pathologie médicale - 4 *vols.*		1	10	0

	£	s.	d.
Richet, A.—Traité pratique d'anatomie médico-chirurgicale -	0	18	0
Robin.—Programme du cours d'histologie - - -	0	6	0
Robin.—Traité du microscope - - - -	1	0	0
———— Anatomie et physiologie cellulaire - - -	0	16	0
———— Leçons sur Les Humeurs, 2ème edition - -	0	18	0
Roubaud.—Traité de l'impuissance et de la stérilité chez l'homme et chez la femme - - -	0	8	0
Saboia.—Accouchements - - - - -	0	13	0
Sandras et Bourguinon.—Traité pratique des maladies nerveuses - - - - *2 vols.*	0	12	0
Saint-Vincent.—Nouvelle médecine des familles à la ville et à la campagne - - - -	0	3	6
Schimper.—Traité de paleontologie végétale *vols.* 1 and 2 *vol.* 3 *sous presse.*	5	0	0
Sedillot et Legouest.—Traité de médecine opératoire 2 *vols.*	1	0	0
Senac.—Traitment des coliques hépatiques - - -	0	4	0
Tardieu. Dictionnaire d'hygiène publique et de salubrité, 4 *vols.*	1	12	0
———— ÉTUDE MÉDICO-LEGALE sur les blessures par imprudence, l'homicide, et les coups involontaires -	0	3	6
———— Ditto ditto sur la pendaison, la strangulation, et la suffocation - - - - -	0	5	0
———— Ditto ditto sur l'avortement - - -	0	4	0
———— Ditto ditto sur l'empoisonnement - - -	0	12	0
———— Ditto ditto sur les attentats aux mœurs - -	0	4	6
———— Ditto ditto sur l'infanticide - - - -	0	6	0
———— Ditto ditto sur la folie - - - -	0	7	0
———— Ditto ditto sur l'identité - - - -	0	3	0
———— Manuel de pathologie et de chirurgie médicales -	0	7	0
Trousseau. — Clinique médicale de l'Hôtel-Dieu de Paris 3 *vols.*	1	12	0
Valleix.—Guide du médecin praticien, résumé général de pathologie et de thérapeutique appliquées, 5 *vols.*	2	10	0
Vandercolme.—Histoire botanique et thérapeutique des salsepareilles - - - -	0	3	6
Vaslin.—Plaies par armes à feu - - -	0	6	0
Vidal.—Traité de pathologie externe et de médecine opératoire 5 *vols.*	2	0	0
Virchow.—Pathologie des tumeurs - - - 3 *vols.*	1	16	0
Vulpian.—Leçons de physiologie générale et comparée du système nerveux au musée d'histoire naturelle -	0	10	0
———— Leçons sur l'appareil vaso-moteur - - *vol.* 1	0	8	0
Wagner.—Traité de chimie industrielle - - 2 *vols.*	1	0	0
Wolliez.—Dictionnaire de diagnostic médical, comprenant le diagnostic raisonné de chaque maladie, leur signes, &c. - - - - - -	0	16	0
Wundt.—Traité élémentaire de physique médicale, traduit de l'Allemand par le Dr. Monoyer - - -	0	12	0
———— Nouveaux éléments de physiologie humaine -	0	10	0
Wunderlich.—De la température dans les maladies -	0	14	0

BIBLIOTHEQUE

DE

PHILOSOPHIE CONTEMPORAINE.

IN VOLUMES 2*s.* 6*d.* EACH.

Alaux.—Philosophie de M. Cousin.
Auber, Ed.—Philosophie de la médecine.
Barot, Odysse.—Lettres sur la philosophie des histoires.
Beauquier.—Philosophie de la musique.
Beaussire.—Antécédents de l'Hégélianisme dans la philos. franç.
Bentham et Grote.—La religion naturelle.
Bersot, Ernest.—Libre philosophie.
Bertauld.—L'ordre social et l'ordre moral.
Buchner, L.—Science et nature.
Bost.—Le protestanisme libéral.
Bouillier (Francisque).—Du plaisir et de la douleur.
—— De la conscience.
Boutmy, E.—Philosophie de l'architecture en Grèce.
Challemel Lacour.—La philosophie individualiste, étude sur Guillaume de Humboldt.
Coignet, C.—La morale indépendante.
Coquerel, Ath.—Origines et transformations du christianisme.
—— La conscience et la foi.
—— Histoire du credo.
Dumont.—Heckel et la Theorie de l'Evolution en Allemagne.
Faivre.—De la variabilité des espèces.
Fontanes.—Le christianisme moderne.   Étude sur Lessing.
Fonvielle, W.—L'astronomie moderne.
Franck, Ad.—Philosophie du droit pénal.
—— Philosophie du droit ecclésiastique.
—— La philosophie mystique en France aux viii* siècle.
Garnier, Ad.—De la morale dans l'antiquité.
Gaukler.—Le Beau et son histoire.
Herzen.—Physiologie de la Volonte.
Janet, Paul.—Le matérialisme contemporain.
—— La crise philosophique.   MM. Taine, Rénan, Vacherot, Littré.
—— Le cerveau et la pensée.
—— Philosophie de la révolution française.'
Laugel, Auguste.—Les problèmes de la nature.
—— Les problèmes de la vie.
—— Les problèmes de l'âme.

Laugel, Auguste.—La voix, l'oreille, et la musique.
—— L'optique et les arts.
Laveleye, Em. de.—Les formes de gouvernement.
Leblais.—Matérialisme et spiritualisme, préface par M. E. Littré.
Lemoine, Albert.—Le vitalisme et l'animisme de Stahl.
—— De la physionomie et de la parole.
Letourneau.—Physiologie des passions.
Levallois, Jules.—Déisme et christianisme.
Leveque, Charles.—Le spiritualisme dans l'art.
—— La science de l'invisible. Étude de psychologie et de théodicée.
Mariano.—La philosophie contemporaine en Italie.
Max-Muller.—La Science des Religions.
Mill, Stuart.—Auguste Comte et la philosophie positive.
Milsand.—L'esthétique anglaise, étude sur John Ruskin.
Moleschott, J.—La circulation de la vie.
Odysse-Barot.—Philosophie de l'histoire.
Remusat, Charles de.—Philosophie religieuse.
Reville, A.—Histoire du dogme de la divinité de Jésus-Christ.
Ribot.—Philosophie de Schopenhauer.
Saigey.—La physique moderne.
Saisset, Emile.—L'âme et la vie, une étude sur l'esthétique franc.
—— Critique et histoire de la philosophie.
Schœbel.—Philosophie de la raison pure.
Selden, Camille.—La musique en Allemagne. Mendelssohn.
Spencer, Herbert.—Classification des sciences.
Taine, H.—Le positivisme anglais, étude sur Stuart Mill.
—— L'idéalisme anglais, étude sur Carlyle.
—— De l'idéal dans l'art.
—— Philosophie de l'art.
—— Philosophie de l'art en Italie.
—— Philosophie de l'art dans les Pays-Bas.
—— Philosophie de l'art en Grèce.
Tissandier.—Des sciences occultes et du spiritisme.
Vacherot, Et.—La science et la conscience.
Vera, A.—Essais de philosophie Hégélienne.

FORMAT IN-8.

	£	s.	d.
Agassiz.—De l'espèce et des classifications	0	5	0
Bain.—Les sens et de l'intelligence			
Barni, Jules.—Le morale dans la démocratie	0	5	0
Quatrefages, de.—Darwin et ses précurseurs français	0	5	0
Saigey, Emile.—Les sciences des 18e siècle	0	5	0
Spencer, Herbert.—Les premiers principes	0	10	0
Mill, Stuart.—La philosophie de Hamilton	0	10	0

BIBLIOTHEQUE
D'HISTOIRE CONTEMPORAINE.

IN VOLUMES AT 3*s.* 6*d.*

Bagehot.—La constitution anglaise.

Barni, Jules.—Histoire des idées morales et politiques en France au XVIIIe siècle. *2 vols.*

———— Les moralistes français des XVIIIe siècle.

———— Napoléon Ier. et son historien M. Thiers.

Barry, Herbert.—La Russie contemporaine, traduit de l'Anglais.

Beaussire, Emile.—La guerre étrangère et la guerre civile.

Boert.—La guerre de 1870-71 d'après Rustow.

Bourloton, Ed.—L'Allemagne contemporaine.

Carlyle.—Histoire de la révolution française.   3 *vols.*

Clamagerau.—La France républicaine.

De Rochau.—Histoire de la restauration.

Despois, Eug.—Le vandalisme révolutionnaire.

Dixon, H.—a Suisse contemporaine, traduit de l'Anglais.

Duvergier, De Hauranne.—La république conservatrice.

Hillebrand.—La Prusse contemporaine et ses institutions.

Laugel, Auguste.—Les États-Unis pendant la guerre (1861-65).

Meunier, Victor.—Science et démocratie.

Montegut, Emile.—Les Pays-Bas.   Impres^{ns} de voyage et d'art.

Reynald, H.—Histoire de l'Espagne depuis la mort de Charles III. jusqu'à nos jours.

Sayous, Edouard.—Histoire des Hongrois et de leur littérature politique de 1790 à 1815.

Teste, Louis.—L'Espagne contemporaine, journal d'un voyageur.

Thackeray.—Les quatre George.

Veron, Eugène.—Histoire de la Prusse depuis la mort de Frédéric II. jusqu'àla bataille de Sadowa.

———— Histoire de L'Allemagne depuis la bataille de Sadowa.

FORMAT IN-8, 7s. EACH.

Alglave, Emile.—Histoire de l'impôt sur le Revenu en France (sous presse).

Delord, Taxile.—Histoire du second empire, 1848-69.   4 vols.

De Sybel.—Histoire de l'Europe pendant la révolution française.   2 *vols.*

Lewis, Sir G. Cornewall.—Histoire gouvernementale de l'Angleterre de 1770 jusqu'à 1830, précédé de la vie de l'auteur.

VALUABLE SCIENTIFIC DICTIONARIES IN COURSE OF PUBLICATION.

	£	s.	d.
Dictionnaire de Chimie pure et appliquée, en livraisons -	0	4	0
Dictionnaire de Médecine, de chirurgie, et d'hygiène vétérinaires. Edition entièrement refondue par A. Zundel, en 6 parties, aux souscripteurs - - - -	2	10	0
Dictionnaire Encyclopedique des Sciences Médicales publié par demi-volume de chacun 400 pages et en trois séries simultanées : la première, commençant par la lettre A ; la deuxième, par la lettre L ; la troisième, par la lettre Q	0	6	6
Nouveau Dictionnaire de Médecine et de chirurgie pratiques, d'environ 30 volumes, chaque - -	0	1	0

Revue des Deux Mondes.   1st and 15th of every Month.

Revue Scientifique de la France et de l'Etranger.   Weekly. 6*d.*

Revue Politique et Literaire.   Weekly.   6*d.*

La Jeune Mere.   Weekly, 6d.

La Petite Bibliographie française *contains a monthly summary of the Literature of the Continent in all its branches.   This small catalogue is published on the 1st of each month, and will be sent gratuitously upon application.*

WORKS NOT IN STOCK, PROCURED AT THE SHORTEST POSSIBLE NOTICE.

Messrs. BAILLIÈRE, TINDALL, & COX *are the specially appointed Agents for the* Revue des Deux Mondes, *and most of the Scientific and Medical Periodicals of the Continent.*